Liquid Crystals with Nano/Micro Particles and Its Applications

Jayeeta Chattopadhyay

Chemistry Department, Amity University Jharkhand, Ranchi
Jharkhand, India

Rohit Srivastava

School of Petroleum Technology, Pandit Deendayal Energy
University, Gandhinagar, Gujarat, India

CRC Press
Taylor & Francis Group
Boca Raton London New York

CRC Press is an imprint of the
Taylor & Francis Group, an **informa** business

A SCIENCE PUBLISHERS BOOK

First edition published 2023
by CRC Press
6000 Broken Sound Parkway NW, Suite 300, Boca Raton, FL 33487-2742

and by CRC Press
4 Park Square, Milton Park, Abingdon, Oxon, OX14 4RN

© 2023 Taylor & Francis Group, LLC

CRC Press is an imprint of Taylor & Francis Group, LLC

Library of Congress Cataloging-in-Publication Data (applied for)

ISBN: 978-0-367-55431-6 (hbk)
ISBN: 978-0-367-55432-3 (pbk)
ISBN: 978-1-003-09352-7 (ebk)

DOI: 10.1201/9781003093527

Typeset in Palatino
by Radiant Productions

Preface

Recent advancements in the field nanotechnology has opened up a new door in the field of liquid crystal. For many materials, the transition between the liquid and the solid phase is not a single-step process, but a range of various mesophases, which are called liquid crystals (LCs). LCs are self-organized anisotropic fluids that are thermodynamically located between the isotropic liquid and the crystalline phase, exhibiting the fluidity of liquids as well as the long-range lattice order that can only be found in crystalline solids. The addition of nanomaterials to a LC material produces a composite or colloidal dispersion, resulting into a revolutionary change in their applications. The use of liquid crystals and materials with special properties, as well as properties at the nanoscale, are discussed. Recent advancements, spanning from liquid crystal-nanoparticle dispersions to nanomaterials generating liquid crystalline phases after surface modification with mesogenic or promesogenic moieties, has been given special emphasis.

Contents

Preface   iii

1. Liquid Crystals and Its Classifications—   1
Fundamental Studies
  1. Introduction   1
  2. What is a Liquid Crystal?   2
  3. Historical Background   3
  4. Classification   5
     4.1 Conventional and Non-conventional LCs   7
     4.2 Phases and Structures of Calamitic LCs   8
  5. Basic Concepts of Ionic LCs   13
  6. Structures of Chiral LCs   19
     6.1 What is Chirality?   19
     6.2 Chiral LCs   20

2. Micro- and Nano-particles Doped Liquid Crystals   34
  1. Introduction   34
  2. Zero-dimensional Nanoparticles   37
     2.1 Metal Nanoparticles   37
     2.2 Quantum Dots   39
     2.3 Magnetic Nanoparticles   40
  3. Quasi-Zero-Dimensional Ferroelectric Nanoparticles   41
  4. One-Dimensional Nanorods   42
  5. One-Dimensional Carbon Nanotubes   46
  6. Two-Dimensional Nanodiscs and Nanoclay Materials   48
  7. Liquid Crystal Phases with Nano-particle Doping   48
     7.1 Zero-dimensional Nano-particle Additives   49
     7.2 One-dimensional Nanoparticle Additives   55
  8. Conclusions   62

3. Potential Applications of Nanoparticles Aided Liquid Crystals   84

   1.   Introduction   84

   2.   Application of Various NPs Aided LCs in Biosensing Techniques   87

   3.   Applications of Gold NPs Doped LC Composites in LCD Systems   93

      3.1   GNPs Capped in Nematic LCs   96

      3.2   GNPs Capped in Cholesteric LCs   103

      3.3   GNPs Capped in Smectic LCs   104

      3.4   Applications of GNPs in PDLC Systems   106

4. Liquid Crystal Nanoparticles in Commercial Drug Delivery System   116

   1.   Introduction   116

   2.   Importance of Liquid Crystals as Nanomaterial   117

   3.   Different Methods and Techniques used for the Synthesis and Characterization of Liquid Crystal Mesophases/Nanoparticles   119

   4.   Applications of Liquid Crystal NPs in Drug Delivery   122

   5.   Factors Affecting Drug Release/Delivery by Lyotropic Liquid Crystals 124

   6.   Conclusions   125

Index   131

CHAPTER 1
Liquid Crystals and Its Classifications
Fundamental Studies

1. Introduction

The liquid crystal (LC) is a scientific domain of study which represents a unique state of materials, which can be characterized by both mobility and order at molecular and supra-molecular state. The potential application of liquid crystal in aerospace domain, molecular biology, micro-electronics and nano-technologies has become one of the most dynamic approach towards the new scientific and technological world. The typical behaviour of liquid crystal emerges under a certain group of conditions, when phases with intermediate characteristics of a 3D ordered solid structures and disordered liquid states are evolved. In a 3D lattice structure, highly structured solids occupy specific sites and indicate their axes in a particular direction (Figure 1.1(a)). Usually, the phases LCs constitute of orientational order, which points along a particular direction, represented by n. There are cases where the positional order has been depicted as 1 or 2 (Figure 1.1(b) and (c)). In the case of isotropic liquid state, the molecules can randomly orient in all possible directions (Figure 1.1(d)).

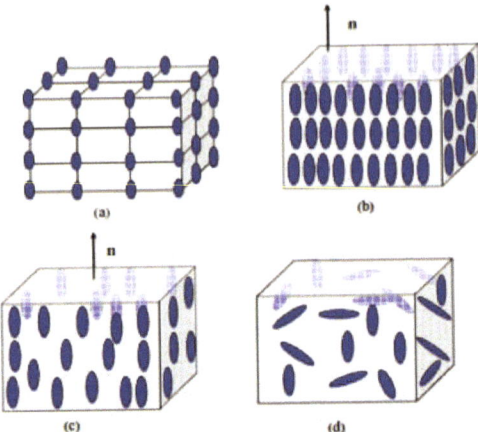

Figure 1.1. The molecular packing in (a) crystals, (b & c) liquid crystals, and (d) liquid state depicted schematically.

In comparison to other solid-state materials, LCs show unique characteristics of their special response towards external stimuli, e.g., light, heat, mechanical forces, surfaces, electric and magnetic fields. They can also remove defects by self-healing [1–4]. Therefore, the intrinsic relationship between chemical structure and their functions of liquid crystalline compounds can be considered. Over the last 130 years, liquid crystals have become one of the most important domains of study in various aspects of chemistry, physics, medicine and engineering with its innovation in structures. The implementation of novel innovative synthetic strategies and characterization technologies have resulted into the development of nanostructured liquid crystallines with special ordering properties. These LCs performed remarkably in electro-optical effect, actuation, chromism, sensing, or templating [5–8]. This chapter will start the historical background of liquid crystal structure, and further it will discuss the basic classification of LCs.

2. What is a Liquid Crystal?

It seems contradictory to define a state of matter in liquid-like and crystalline simultaneously. But they indeed can possess both properties. The liquid crystalline mesophases can possess properties of a liquid, like fluidity and inability to support shear, generation and coalescence of droplets. At the same time, they carry certain properties of crystalline structure, like anisotropy of optical, electrical and magnetic properties, and also a periodic arrangement of molecules in more than one spatial direction. Based on the arrangement of the molecules in a mesophase or its symmetry, LCs can be subdivided into many groups, like, nematics, cholesterics, smectics and columnar mesophases. We will discuss these sub-classes of LCs in the later section of this chapter.

3. Historical Background

The liquid crystalline structure was discovered over 100 years ago in 1888 by Reinitzer [9] and Lehmann [10] when they were working on some esters of cholesterol. In the year 1888, F. Reinitzer, an Austrian botanist and chemistry working at the University of Graz, synthesized several esters of cholesterol, which occur in plant and animal bodies. He observed the phenomenon of "double melting" in these esters. The term "double melting" means the transformation of the compound from crystalline solid phase to an opaque liquid at a certain temperature, and thereafter it changes into an optically clear liquid at a certain higher point of temperature. In various compounds, these phase transformations were reproduced with raising and diminishing temperature. It is clear from the literature that, before Reinitzer, many scientists dealt with liquid crystals, however they did not find this phenomenon with proper minuteness, therefore could not become aware of this unique state. Reinitzer also could not explain properly the curious phenomenon of "double melting" and the simultaneous existence of opaque liquid state. Therefore, he referred these to Otto Lehmann, Professor of Physics at Technical High School of Karlsruhe and leading crystallographer in Germany. He observed typical optical anisotropic behaviour of the cholesterol esters and intuitively affirmed the unique nature of these liquids due to the presence of elongated molecules oriented parallel with the long axes. Later on, it became clear that he explained the phenomenon in the right way. He further designated the term "fluid crystals" ("flieBende Kristalle") and "liquid crystals" ("fliissige Kristalle"). He and other scientists had categorically defined "thermotropic" LCs, as the compounds with liquid crystalline state during certain temperature intervals. On the other hand, "lyotropic" liquid crystals were defined as the mixtures of certain organic salts and water, with various structures of liquid crystals, which only can exist in mixtures of these kinds only [11]. After 1900s, D. Vorländer, who was a Professor in Chemistry, University Halle, started synthesis process to establish intrinsic relation between molecular structure of a compound, with the occurrence of liquid crystalline state in that. Finally, in the year 1908, he depicted that, liquid crystalline compounds must have a molecular shape as linear as possible. Till his retirement in the year 1935, he synthesized around 1100 liquid crystalline substances, 90% of them were known until 1960. He detected the existence of liquid crystalline polymorphism in the year 1906, in which a given compound can exhibit more than one crystalline phase. Vorländer was acclaimed as "father of the liquid crystal chemistry". The researchers from France were very much active in liquid crystal research in the first three decades of this century. One of those scientists, G. Friedel from University of Strasbourg established specific structures of these phases by taking pictures in the polarizing microscope, and it was performed before investigations on liquid crystalline state through X-ray. In the year 1949, Onsagar established a theoretical explanation of these liquid crystalline states based on a model of long stiff rods. This model proved the dominant role of the repulsive forces. In the year 1949, Marier and Saupe reported the importance of the dispersion forces by considering molecular statistical data. The research work on liquid crystals became highly popular among researchers in the mid-1960s due to their increasing applications in optoelectronic displays and for thermography. In 1977, S. Chandrasekhar reported the ability of disc-like molecules to form liquid crystalline structures, this was not restricted only to rod-like molecules [12]. Later, in 1994, Blinov et al. had reduced these gaps between rod and disc like structures by developing lath-like molecules with ability to form liquid crystalline phases [13].

4. Classification

The various states of liquid crystal can be generated either by application of heat energy on mesogens or by solvent treatment on amphiphilic systems. The mesophases generated by temperature variation are termed as thermotropic. On the other hand, mesophases which behave thermodynamically stable with increase or decrease of temperature are called enantiotropic. Although, the thermotropic mesophases which can appear only on cooling are considered as monotropic. Similarly, liquid crystalline phases generated by dissolving the compound in a concerned solvent with appropriate concentration and temperature conditions, are known as lyotropic. When both heat treatment and solvent effect generate various liquid crystalline phases, they are termed as amphotropic. In our daily life, lyotropic LCs are more significant, moreover, life is critically supported by such ordered fluids. However, thermotropic LCs have drawn greater attention due to their simplicity to handle and they can easily serve as a significant medium in fabricating low-power display devices.

The categorization of mesogens and mesophases has always been a complicated task. Over the last two decades, a rich variety of mesophases have been identified through conventional or non-conventional or new molecular structure. Usually, few basic principles are followed to categorize them. Primarily they are classified as being lyotropic or thermotropic, by categorizing their method of realization, which leads to different significant mesomorphism. The basic principle behind the mesophase formation is completely different in thermotropics and lyotropics. In the case of thermotropics, the mesophase forms through the organization of individual molecules, whereas in the case of lyotropics, the solute constituent molecules get aggregated initially, and these structures then form different mesophases depending on temperature and concentration [14, 15]. Furthermore, these materials can be classified in a variety of ways, including structure (as amphiphilic and nonamphiphilic molecules), molecular shape (calamitic, discotic, and banana mesogens, for example), molecular size (as low- and high-molecular-weight compounds), and the type

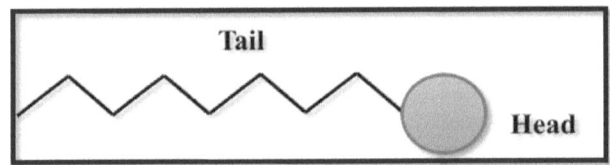

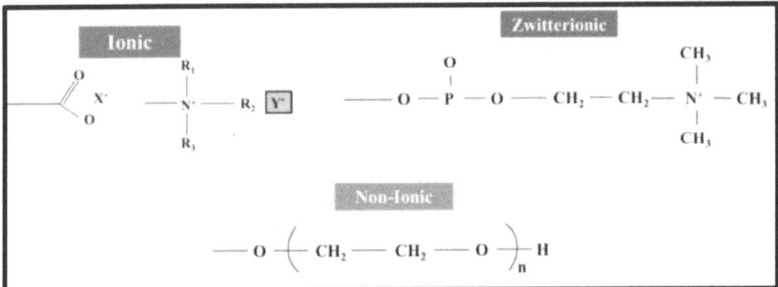

Figure 1.2. Amphiphilic molecules constituted with a polar hydrophilic head and a hydrophobic tail group.

of mesophase formed (nematic, cholesteric, smectic, columnar, and cubic mesophase, etc.).

a) **Lyotropic LCs:** A flexible lipophilic chain (the tail) and a polar (ionic or non-ionic) head group are commonly found in compounds that produce lyotropic mesophases. In most cases, the tail is an alkyl chain with 6 to 20 methylene groups. The basic structures of head groups are represented in Figure 1.2.

Different mesophases can be identified depending on the molecular structure, solvent, amphiphile concentration in the solvent, and temperature. The separation of incompatible (hydrophilic polar and hydrophobic non-polar) components of individual molecules drives them to form Lyotropic LCs. Actual molecular mixtures arise when amphiphilic molecules, such as surfactants, are introduced to a polar solvent at low surfactant concentrations. After reaching a threshold concentration, they form micelles, which are tiny aggregates with a finite size that allow the polar groups to occupy the interface with the polar solvent. It has a spherical shape and is typically the size of a few molecular lengths. Micelles can transform into disc-like, cylindrical, and plate-like supramolecular aggregates when the surfactant concentration is increased further, generating distinct nematic, cubic, hexagonal columnar, and lamellar lyotropic mesophases. The formation of a lyotropic phase can be exhibited by soap dissolving in water. It can be found in all living things, with biological membranes, DNA, and other examples. Apart from its usefulness in biological systems, lyotropic mesophases are also of great interest, as evidenced by the finding of a biaxial nematic phase in a lyotropic system for the first time [16].

b) **Thermotropic LCs:** When crystals are heated, they lose their long-range positional and orientational orders, resulting in an isotropic liquid phase. If the molecules have a certain level of shape anisotropy, the long-range translational periodicity in the crystal may disintegrate in one, two, or three dimensions before the long-range orientational order collapses. Such compounds do not exhibit a single transition from solid to liquid, but rather a series of transitions involving LC phases with mechanical and symmetrical properties that are halfway between liquid and crystal. Melting point refers to the temperature at which a crystal turns into mesophase, whereas clearing point refers to the temperature at which it transitions from mesophase to isotropic condition.

4.1 Conventional and Non-conventional LCs

Conventional LCs: Calamitics and discotics, respectively, are rod-like and disc-shaped mesogens which exhibit thermotropic mesomorphism. For many years, it has been assumed that the molecule, such as 4-methoxybenzylidene-4'-n-butylaniline, MBBA, has to be long or rod-shaped (I) (Figure 1.3). The rigid cores, RC1 and RC2, are aromatic in nature (e.g., 1, 4-phenyl, 2, 5-pyrimidinyl, 2, 6-naphthyl, etc.) but can also be alicyclic (e.g., trans 4-cyclohexyl,

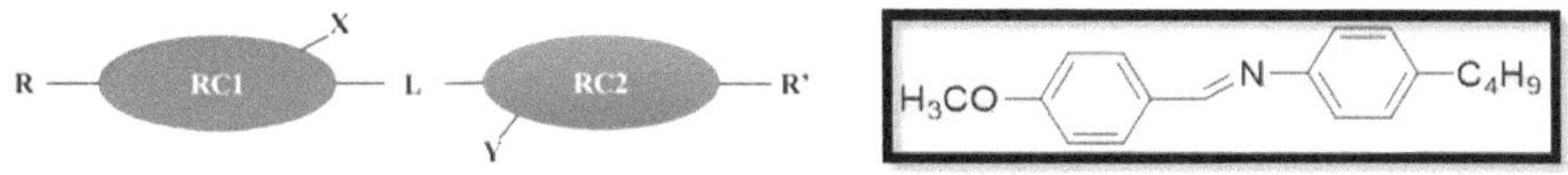

Figure 1.3. General templates of rod-like mesogens.

cholesteryl, etc.). In many situations, these two cores are linked together by a covalent connection, whereas in others, they are linked together by linking unit L (e.g., -COO-, -CH$_2$-CH$_2$-, -CH=N-, -N=N-, etc.). R and R' are common alkyl or alkoxy chains at the end of the thread. One of the terminal units is often a polar substituent (e.g., CN, F, NCO, NCS, NO$_2$, etc.). The lateral units X and Y (e.g., F, Cl, CN, CH$_3$, etc.) are incorporated into the main molecular structure in some exceptional instances.

Non-conventional LCs: The development of self-organized systems with complex mesophase morphologies has received a lot of interest in recent years. This is accomplished by adjusting the volume percentages of incompatible segments or increasing the number of incompatible units in the molecules to modify the shape of stiff segments. Non-conventional liquid crystals are molecules that have an anisometric shape which differs from the standard rod or disc shape [17].

4.2 Phases and Structures of Calamitic LCs

Friedel et al. [18] originally categorised calamitic mesogens into three categories of mesophases: nematic, cholesteric, and smectic, based on the degree of positional and orientational order. The appearance of new phases in liquid crystals is frequently linked to an increase in complexity. The creation of chirality-induced helical supermolecular structures like TGB and Blue phases are two examples.

4.2.1 Nematics

Molecules in a nematic mesophase have a long-range orientational order, with their long axes aligned in a preferred direction. The placements of the centres of mass of molecules have no long-range order. The chosen direction, which varies depending on the media, is referred to as a director. A unit vector, n, is used to denote the director's orientation (r). The molecules in a nematic can spin along their long axes, and their ends, even if they differ, have no favoured arrangement. As a result, the director's sign has no physical meaning, and the nematic behaves optically as an uniaxial material

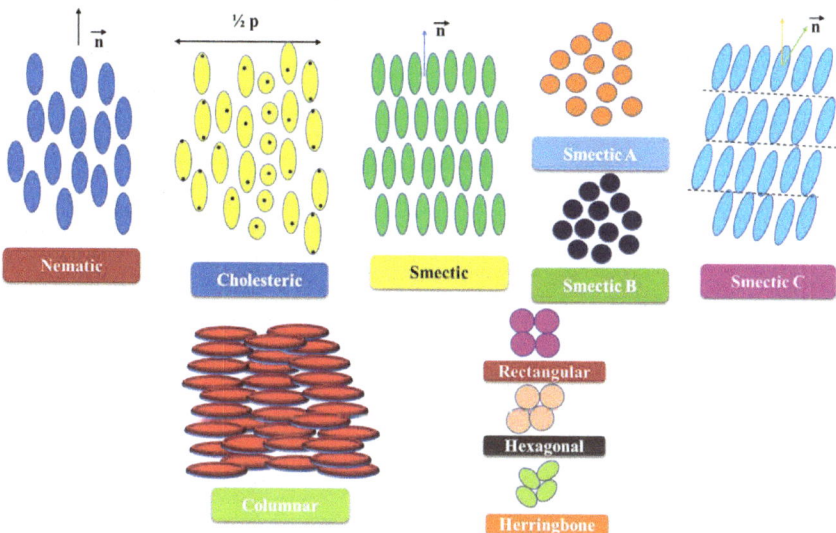

Figure 1.4. Molecular arrangements in various liquid crystalline mesophases: Nematic mesophase-molecular orientations are correlated, not molecular locations. The most common orientation is referred to as a director, n. The average molecular orientation twists across the medium with a specific regularity in a Cholesteric mesophase, although the positions of molecules are not correlated. The molecules in a Smectic A mesophase are arranged in planes. The axes of the molecules are perpendicular to these planes, but they are not ordered within them. In Smectic B, molecules are packed hexagonally in the planes, but in Smectic C, the director is slanted in the planes. Disc-shaped molecules frequently form Columnar mesophases. Hexagonal, Rectangular, and Herringbone are the most prevalent column layouts in two-dimensional lattices.

with a symmetry centre. In Figure 1.4, the director and molecular arrangement in a nematic mesophase is represented, with ellipses representing the anisotropic form of molecules. We rarely observe the desired uniform equilibrium arrangement of the director while optically inspecting a nematic mesophase. A schlieren texture of a nematic is shown in Figure 1.5(a), which was captured using a microscope with crossed polarizers. Every point-defect has four dark brushes indicating that the director is parallel to the polarizer or analyzer. The colours are Newton colours for thin films and vary depending on the sample thickness. Since point defects can only occur in pairs, there are two types of defects: the first has yellow

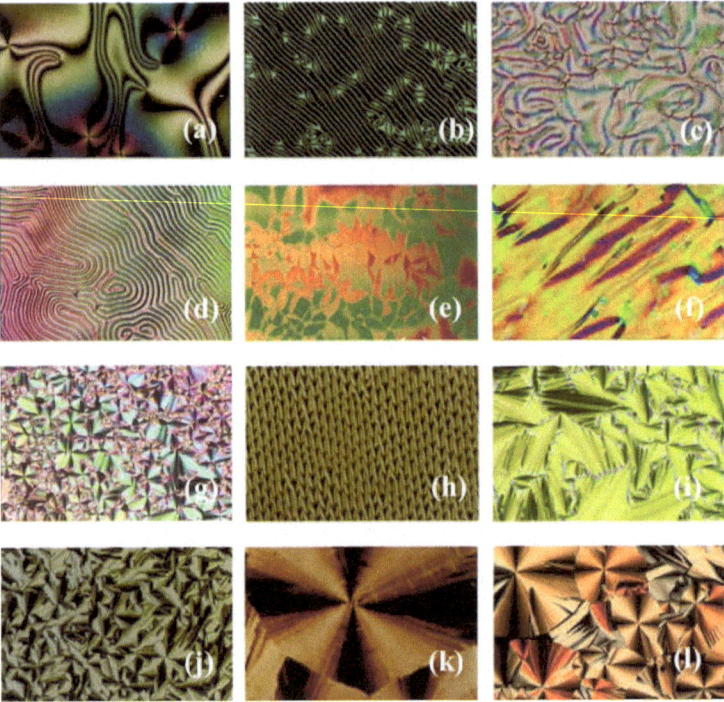

Figure 1.5. Artwork on Liquid crystalline structure: Nematics: (a) Nematic film with surface point defects of Schlieren texture. (b) Thin nematic film on an isotropic surface. (c) Thread-like texture of Nematic. Nematic, the Greek word for "thread" found on these textures. (d) Cholesteric fingerprint texture: The line pattern formed due to the presence of helical structure of the cholesteric phase, with the helical axis present in the plane of the substrate. (e) A short-pitch cholesteric liquid crystal in Grandjean or standing helix texture, observed between crossed polarizers. The bright colors produced due to the differentiation in rotatory power appearing from domains with different cholesteric pitches. This pattern takes place upon rapid cooling close to the smectic A* phase, where the pitch strongly diverges with diminishing temperature. (f) Long-range orientation of cholesteric liquid crystalline DNA mesophases in a magnetic field. (g, h) Focal conic texture of a chiral smectic A liquid crystal. (i) A Chiral smectic C liquid crystal with focal conic texture. (j) Hexagonal columnar phase with a typical spherulitic texture. (k) A discotic liquid crystal with rectangular phase. (l) Liquid-crystalline phase with hexagonal columnar structure. Source: Photos courtesy of Oleg Lavrentovich (http://www.lavrentovichgroup. com/textures.html), Ingo Dierking (http://softmatter-_dierking.myfreesites.net), Per Rudqvist, Sivaramakrishna Chandrasekhar, Prasad Krishna, Nair Gita, and Jon Rourke (https://www2.warwick.ac.uk/fac/sci/chemistry/research/rourke/ rourkegroup/mesogens/).

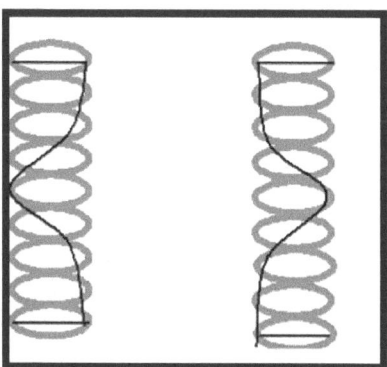

Figure 1.6. Schematic illustration of a left (right part) and a right handed (left part) cholesteric helical superstructure.

and red brushes, while the second is less colourful. A shot of a thin nematic layer on an isotropic surface is used to create the texture in Figure 1.5(b). In a restricted film, the periodic stripes result from a conflict between bulk elastic and surface anchoring forces. Surface anchoring aligns molecules perpendicular to the top surface of the film and parallel to the bottom surface. The director field's distortions are fought by elastic forces. The lowest energy state corresponds to the director in the plane of the film when the film is suitably thin. A thread-like texture may be seen in the pattern in Figure 1.5(c). Threads are disclinations, which are equivalent to dislocations in solids. These Nematics draw their name: "μγλα", which is a Greek word and means of "thread". Chemically, molecules that form nematic crystalline mesophases have anisotropic structures, with hard molecular backbones defining the long axes of molecules. Molecules with flat segments, such as benzene rings, are more prone to form liquid crystallinity. The dipole moments of many liquid crystalline compounds are high, and the groups are easily polarizable. Figure 1.6 depicts typical chemicals that form nematics.

4.2.2 Cholesterics

The cholesteric mesophase resembles the nematic since it has a long-range orientational order but no long-range positional order of molecular centres of mass. It differs from the nematic mesophase as the director varies in a predictable pattern across the medium, even when it is unstrained. The director distribution is exactly what you'd get if you twisted a nematic along the y axis around the x axis. The long axes of the molecules align along a single preferred direction in any plane perpendicular to the twist axis in this plane, but this direction rotates evenly through a sequence of parallel planes, as seen in Figure 1.4. The distance measured along the twist axis over which the director revolves in a full circle defines the secondary structure of the cholesteric. The pitch of the cholesteric, p, is the distance between the two points. Because n and n are indistinguishable, the cholesteric periodicity length is really half of this distance. Formally, a nematic liquid crystal is an infinite pitch cholesteric. As a result, there is no phase transition between the nematic and cholesteric mesophases: nematics doped with enantiomorphic chemicals become lengthy but finite-pitch cholesterics. As seen in Figure 1.6, the molecules that make up the cholesteric mesophase exhibit different right- and left handed configurations. Common cholesterics have a pitch of many hundreds of nanometers, which is comparable to visible light wavelength. The periodic spiral arrangement is responsible for the unique hues of cholesterics in reflection, as well as their extremely high rotatory power, due to Bragg reflection. Temperature, flow, chemical composition, and applied magnetic or electric fields can all affect pitch [19]. Figure 1.5 depicts typical cholesteric textures.

4.2.3 Smectics

The stratification of a smectic mesophase is a key characteristic that distinguishes it from a nematic or cholesteric one. In addition to the directional ordering, the molecules are organized in layers and show certain linkages in their positions. The layers are free to slip over one another. Various sorts of smectics have been observed depending on the molecular order in layers. As seen in Figure 1.3, molecules in smectic A are arranged perpendicular to the layers, with no long-range crystalline ordering inside them. The preferred molecular axis in smectic C is not perpendicular to the layers, resulting in biaxial symmetry. Within the layers of smectic B, there is a hexagonal crystalline arrangement. Smectics do not assume the basic structure seen in Figure 1.3 when put between glass substrates. In order to fit the substrates, the layers become deformed and can slide over one another to maintain their thickness. These aberrations give rise to the smectic focal conic texture. Typical smectics textures are depicted in Figure 1.4. Nematic (or cholesteric) and smectic mesophases exist in a number of substances. The lower temperature phases, on an average, have a higher degree of crystalline organisation. The nematic mesophase always occurs at a higher temperature than the smectic one; as the temperature drops, the smectic mesophases occur in the following order: A → C → B.

4.2.4 *Columnar Mesophases*

Columnar mesophases are a type of liquid-crystalline phase in which molecules form cylindrical formations. Because the columnar formations are made up of stacked flat-shaped discotic molecules, such as triphenylene derivatives illustrated in Figure 1.6, these liquid crystals were originally dubbed discotic liquid crystals. Recent findings have shown a variety of columnar liquid crystals made up of non-discoid mesogens, which are now known as columnar liquid crystals [20]. The packing motivation of the columns is used to organise columnar liquid crystals. Molecules in columnar nematics, for example, do not form columnar assemblages and instead float with their short axes parallel to one another. Columns are stacked in two-dimensional lattices in various columnar liquid crystals, such as hexagonal, tetragonal, rectangular, and herringbone, as seen in Figure 1.4.

5. Basic Concepts of Ionic LCs

Ionic liquid crystals are materials that combine some of the qualities of liquid crystals (such as anisotropy of physical properties) with those of ionic liquids (such as ionic conductivity). The massive rise in research activity on the latter over the last two decades has undoubtedly aided the growing interest in ionic mesogens. This is demonstrated by the fact that two of the most frequently cited articles on ionic liquid crystals [21–22] are from well-known ionic liquid crystals research groups (not liquid crystals). Ionic liquids (ILs) are solvents that are fully made up of ions that melt at temperatures below 100 degrees Celsius [23–26]. The electrochemical application of ILs as electrolytes was the focus of early research in the 1970s and 1980s. Moisture-sensitive halogenoaluminate (III) salts were used in the "first-generation" ILs. The fact that aprotic molten salts have an exceptionally low vapour pressure, making them ideal for substituting volatile organic solvents in organic synthesis and allowing their usage in high-vacuum systems sparked more interest in ILs, particularly room-temperature ILs [27–29]. Many ILs also have a wide liquidus range, excellent thermal stability, and/or a wide electrochemical window, and they are frequently non flammable. In 1992, 1-ethyl-3-methylimidazoliumtetrafluoroborate ([C2mim][BF4]; [C2mim]$^+$ = 1-ethyl-3-methylimidazolium) was reported as the first air- and moisture-stable "second-generation" IL [30]. Since then, ILs containing an organic cation and an "inert" anion such as [BF$_4$], [PF$_6$], [OTf]$^-$, or [NTf$_2$]$^-$ have mainly replaced halogenoaluminate (III) salts. It should be noted, however, that [PF$_6$] and especially [BF$_4$] anions can slowly hydrolyze to HF in the presence of water [31–34]. Because a material with certain desired qualities (e.g., (im) miscibility with water and other solvents, dissolving ability, polarity, viscosity, density, etc.) can be made by combining a suitable cation and an appropriate anion, ILs might be termed "designer solvents".

Because most ILs are intrinsically ion-conducting fluids, they've been studied as (a) reaction media for organic reactions, where higher catalytic activity and/or selectivity have been observed; (b) solvents for immobilising transition metal catalysts in biphasic catalysis; (c) extraction solvents and gas processing media; and (d) electrolyte solution substituents for batteries, fuel cells, solar cells, and capacitors; (e) as electrodeposition media for reactive metals and semiconductors; (f) as luminescence solvents; and (g) as templating agents for the creation of inorganic nanostructures [35–52]. They've already proven useful in the chemical industry [53]. Since ILs mostly constitute of ions with poor coordination, they have the potential to be highly polar yet noncoordinating solvents. The use of ILs in the manufacture of inorganic materials is a new field that is rapidly gaining traction. Because of the nonvolatility and thermal stability of many ILs, water (and other non ionic solvents) may be practically completely removed, and hitherto unknown inorganic compounds can be produced by using the IL as a "structure-directing" (templating) solvent [53–60]. The generation of ("lyotropic") LC mesophases is being studied using both protic and nonprotic ILs as amphiphile self-assembly medium [60–72].

It's crucial to understand the difference between "ionic liquid" and "ionic liquid crystal." The phrase "ionic liquid crystal" refers to an ionic liquid-crystalline material that contains cations and anions and has at least one liquid crystalline mesophase (either enantiotropic or monotropic). It's a slang term for "ionic mesogen." The phrase can be applied to any material, regardless of its melting point (which does not have to be below 100°C), viscosity (which can be quite high), or potential use as a reaction medium. Even in their "isotropic liquid" condition, many non mesomorphic ILs exhibit nanoscale architecture, charge ordering, and local anisotropy. SAXS/WAXD and neutron scattering measurements, as well as other experimental techniques and molecular dynamics (MD) simulations, show that typical ILs with sufficiently long alkyl chain substituents (already for propyl or butyl side chains in the case of imidazolium ILs) have some spatial mesoscopic structural heterogeneity in their liquid state [73–97]. Aromatic ILs and nonaromatic ILs have been found to be different from each other [98]. The heterogeneity is due to polar headgroup aggregation (or network formation) (due to the charge-ordering effect resulting from strong long-range coulombic interactions, but cooperative hydrogen-bonding interactions between the cations and anions, which induce structural directionality, also play a significant role) and concomitant domain formation by the alkyl chains, though not to the same extent or on the same length scale as ILCs. Longer alkyl chains (typically at least undecyl or dodecyl) are present in the latter, allowing for true long-range correlated microphase segregation with long-range positional order/periodicity [99]. Nonetheless, $[C_n \text{ mim}]^+$ (n > 11) ILCs show a logical shift from the close-packed cationcation second-shell radial distribution seen in $[C_1 \text{ mim}]^+$ salts to the smectic layers with interdigitating alkyl chains seen in $[C_n \text{ mim}]^+$ salts [100]. Bradley et al. and De Roche et al. argued that some short-range structural ordering, likely due to surviving ionic aggregates or clusters, is still present in the isotropic liquid phase of the latter compounds based on SAXS and ionic conductivity measurements[101,102]. The phenomenon was confirmed by adiabatic scanning calorimetry (ASC) measurements on piperidinium- and morpholinium-based ILCs. Abdallah et al. also reported persistent ordering in relation to phosphonium-based ILCs [103], and the phenomenon was confirmed by adiabatic scanning calorimetry (ASC) measurements on piperidinium- and morpholinium-based ILCs [104]. For typical ILs with very short alkyl chains, such as [C_2mim] [NO_3], these nanoscale spatial heterogeneities are significantly less prominent and only last a brief time [100, 105, 106]. In contrast to their isoelectronic alkyl-substituted counterparts, ILs with short to medium-sized ether-containing alkyl substituents showed a similar lack of liquid-state spatial heterogeneity, both theoretically and empirically [107–109].

While most liquid crystals (LCs) are neutral organic molecules, there are also many ionic liquid crystals (ILCs). Mesogens are substances that include both (positively charged) cations and (negatively charged) anions. The properties of ILCs may differ dramatically from those of regular LCs due to their ionic nature. Since the 1980s, there has been an increase in the number of reports in the literature on various forms of thermotropic ILCs [110–112]. Ammonium salts and pyridinium salts, as well as metal carboxylate complexes, were studied [113–117]. Amphiphilic pyridinium salts have been known to have thermotropic mesomorphism since 1938 [118]. Other organic cations, such as substituted imidazolium, phosphonium, and viologen (1,1'-disubstituted 4,4'-bipyridinium) cores, were gradually added to the mix to produce mesomorphic salts. Apart from large positively charged fused ring systems and protonated cyclo[8]pyrroles, ILCs are usually based on pyrrolidinium, piperidinium, morpholinium, piperazinium, -caprolactam, guanidinium, triazolium, pyrazolium, benzobis-(imidazolium), quinolinium, isoquinolinium, pyrimidinium, 1,10-phenanthrolinium systems (Scheme 1.1). The quaternary ammonium, pyridinium, imidazolium, phosphonium, and viologen ILCs sparked interest due to their ease of synthesis by quaternization with alkyl halides. Because the cationic half of ILCs

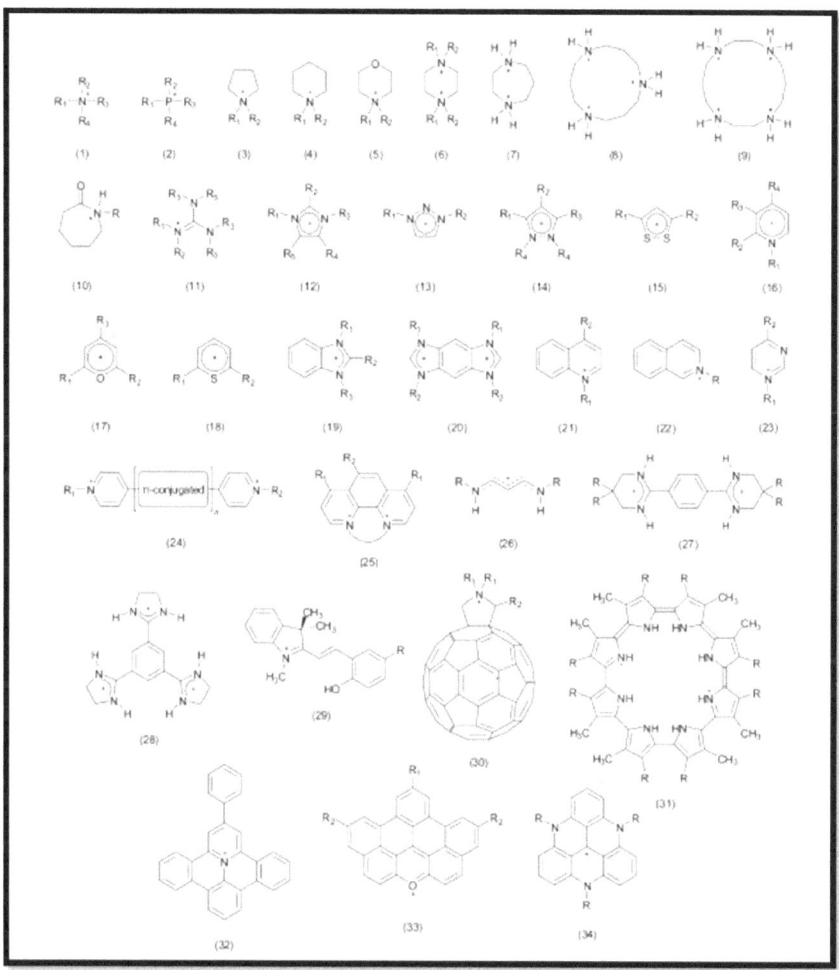

Scheme 1.1. Organic cations which have been used to make low molar mass thermotropic ILCs: general molecular structure.

is often easier to synthesise and substitute than the anionic part, it is common practise to categorise the different forms of ILCs according to the type of cation. As a result, the various sections of this review have been designed accordingly. There are cases of "anion-induced" mesomorphism. There are various review articles that (partially) address the area of ILCs, the most complete of which is Binnemans' review from 2005 [113–116, 119–122]. A further distinction can be made between (i) ILCs with cations carrying simple long alkyl chains (which account for the majority of all investigated compounds, owing to their ease of synthesis); (ii) ILCs with the charged moiety (e.g., an imidazolium cation, which is a polarizable rigid group) incorporated in the rigid part of the mesogen; and (iii) ILCs with conventional mesogenic groups attached via the strong amphiphilic character of compounds of type (i) is primarily responsible for their thermotropic LC characteristics, which also explains their lyotropic mesomorphism [123, 124]. Apart from amphiphilicity, compounds of type (ii) are instances of more traditionally structured systems, in which excluded volume effects play a significant role. Nonsymmetric LC dimers (two stiff moieties whose molecular mobility is constrained due to their link via a flexible spacer) can be thought of as compounds of type (iii) [125–127]. Type (iv) compounds with a wedge shape can form cylindrical supramolecular aggregates that produce columnar and cubic mesophases. The combination of amphiphilic compounds with anisometric structural units bridges the gap between thermotropic and lyotropic LCs [128].

The utilisation of the so-called ionic self-assembly (ISA) idea to generate useful nanostructured materials, such as ILCs, is a relatively new trend in materials research [129–133]. In essence, this technique entails combining cations and anions (usually charged oligoelectrolytes and oppositely charged surfactants) to construct highly ordered, hierarchical superstructures, with noncovalent electrostatic interactions serving as the fundamental driving force. Hydrophobic interactions, interactions, hydrogen bonding, and other secondary driving mechanisms for self-organization exist. ISA is related to molecular tectonics, which is the assembly of complementary molecular building blocks (= "tectons") into predesigned 1D, 2D, or 3D periodic architectures via all kinds of noncovalent interactions (this term is commonly used in the context of self-assembly in crystalline solid phases or on surfaces) [134]. Combining oppositely charged, functional, structurally different, and preferably simple (or commercially available) oligoelectrolytic building blocks is a simple (or commercially available) way to obtain processable nanostructured materials with specific properties without the use of complicated covalent chemistry. The ISA concept is not always straightforward to put into practise since proper reaction conditions for the actual ion exchange must be identified (stoichiometric precipitation of high-purity products). True ISA, on the other hand, is accompanied by a cooperative binding process, in which the initial bonds encourage subsequent binding of charged molecules toward a final self-assembled structure [129, 130].

6. Structures of Chiral LCs

6.1 What is Chirality?

Chirality [1–4] is linked to symmetry, or more accurately, the lack of it. Symmetry breaking occurs in all elements of nature, whether living or non-living, huge or microscopic, molecule or macroscopic. Since Louis Pasteur's discovery that chiral crystals must be composed of asymmetric molecules [5], the investigation of chirality consequences has attracted scholarly and industrial interest. The term chirality, originating from the Greek word κειρ (hand), is now widely recognised as a very standard phenomenon. The absence of mirror symmetry is fairly prevalent, with the most notable case being human hands, which cannot be drawn together by using symmetry elements of the first kind, translations and rotations. They are mirror images of each other and, as may be shown by putting a right glove on a left hand. Chirality is frequently observed in molecules, in which two mirror counterparts (enantiomers) have comparable fundamental qualities, such as melting temperature, but chiral characteristics of different handedness, causing dramatic alterations. D-limonene, for example, has an orange aroma, but L-limonene has a lemon aroma. In the formulation and function of pharmaceuticals, such modest chirality effects are frequently used [7–9]. Chirality, on the other hand, can be found in all sizes. The emission of nearly solely single handed electrons (right handed) and neutrinos (left handed) during the beta decay of Cobalt, Co60, has been demonstrated to exhibit handedness, with the electron moving in the direction of the spin (termed parity violation caused led by the weak interaction). This has been proposed as a possible explanation for the handedness of physiologically active compounds like amino acids and carbohydrates. On the other hand, it appears that galaxies have a certain handedness [10–12], which makes spiral galaxies appear two-dimensional when projected onto a plane. The northern sky appears to have more left-handed spirals, whereas the southern sky appears to have more right-handed spirals.

An asymmetry of molecular groups, most commonly an asymmetrical substituted carbon atom, causes molecular chirality. This is referred to as a chiral centre. Chiral axes and planes are used less frequently to introduce chirality. Cahn, Ingold, and Prelog (CIP-system) [13] presented a set of guidelines for determining left and right handedness. On the other hand, defining a chirality measure is not easy [14]. Optical activity, circular dichroism (CD), and optical rotation dispersion are examples of variables that can be used to compare two chiral substances (molecules) (ORD).

6.2 Chiral LCs

Like other compounds, liquid crystal molecules (mesogens) can be made chiral [135]. There are two basic approaches: (i) incorporating chirality inside the molecule, primarily by asymmetrically swapping one or more carbon atoms with four distinct ligands. There are also examples of mesogens having a chiral axis or plane, but they are far less common; (ii) adding mesogenic or non-mesogenic chiral dopants to an existing liquid crystal phase at various concentrations. In general, chirality-related effects grow linearly with dopant concentrations below 5%–10% by weight for modest concentrations below 5%–10% by weight. Phase separation is common in higher volumes, especially if the mesogen and chiral dopant have different molecular shapes. The addition of chirality to the system has a significant impact on the characteristics, just as it does for other compounds. Several phases, such as the chiral nematic or cholesteric phases, have helical superstructures, in which the director describes a spiral in space along the z-direction according to Equation (i), where P is the pitch of the helix and 0 is a constant that depends on the boundary conditions:

$$n(r) = \begin{pmatrix} \cos\left(\dfrac{2\pi}{P}z + \varphi_0\right) \\ \sin\left(\dfrac{2\pi}{P}z + \varphi_0\right) \\ 0 \end{pmatrix} \tag{i}$$

6.2.1 Chiral Nematic Phase

The helicity of the cholesteric or chiral nematic phase is one of the most visible expressions of macroscopic symmetry breaking through molecular chirality. In this scenario, the director twists around an axis that is perpendicular to the long molecular axis, forming a spiralling shape. The left-handed helix is produced by one enantiomer, whereas the right-handed helix is produced by its mirror copy (Figure 1.6). This is specified by attributing a sign to the pitch P, which is usually negative for a left-handed helical superstructure and positive for a right-handed helical superstructure. Depending on the boundary conditions, the pitch can be depicted in a variety of ways. With the helical axis perpendicular to the substrate, a so-called Grandjean texture is observed, whereas with homeotropic boundary conditions, an equidistant line pattern allows the pitch to be measured, as the former has a periodicity of P/2 [136].

The temperature dependency of the cholesteric pitch is such that as the temperature rises, the pitch decreases. This seemingly paradoxical effect can be explained descriptively by a rotating equivalent of thermal expansion. Because of their chirality, the molecules perform torsional vibrations in an anharmonic potential. As the temperature rises, the angle of these vibrations increases, and the helix's pitch decreases [137]. Intermolecular interactions produce the anharmonic potential and, as a result, the pitch's temperature dependence. Figure 1.7 displays an example of a N*-SmA* and N*-SmC* transition. The cholesteric pitch separates as it approaches the smectic phase, because its twist is incompatible with a layered structure.

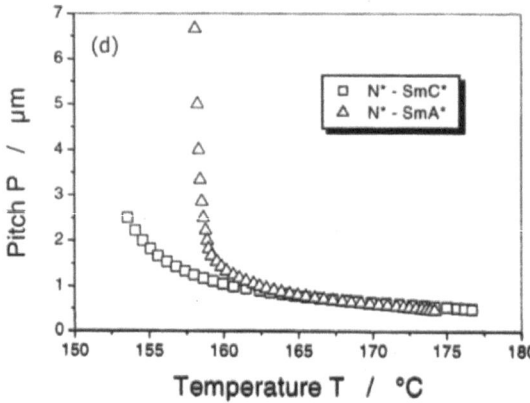

Figure 1.7. Temperature dependence of the supermolecular helix approaching the N*-SmA* and N*-SmC* transition [Reproduced from Ref. 136].

Pitch measurements as a function of temperature enable the identification of critical exponents, which should be different for the transition into the SmA* and SmC* phases, according to predictions. These views have been the subject of a long-running discussion. Three potential theoretical concepts have been included here: (a) $v = 1/2$ for N*-SmA* and $v = 3/4$ for N*-SmC* [138], (b) $v = 1/2$ for N*-SmA* and $v = 1$ for N*-SmC* [139] and $v = 1/2$ for N*-SmC* [140]. However, the experimental evidence was quite debatable. Evidence for the Chen and Lubensky model has recently been provided with a critical exponent of $v = 1/2$ for the cholesteric to smectic A* transition and $v = 1$ for the cholesteric to smectic C* pitch divergence for cholesteric phases with a wide temperature range and proposing a sensible temperature dependence of the pitch, which combines a natural pitch dependence in superposition with pretransitional effects for cholesteric phases with a wide range [141]. This means that variations must be taken into account while discussing these transitions. Another fascinating behaviour, which is not so common, is termed as cholesteric twist inversion [142, 143]. While this is easily explained in combinations of different components due to the pitch's temperature dependence [144], it is far more startling in a single component system. Temperature alters the handedness of the helical superstructure, causing it to traverse through a nematic director configuration. At first view, this seems to be in violation of Louis Pasteur's rules. Despite this, each of these molecules has at least two chiral components, most commonly chiral centres. The cholesteric twist inversion of single component systems has been demonstrated to be described by a superposition of each chiral center's contributions [145].

6.2.2 Lyotropic LC Phases in Cellulose

Cellulose and its derivatives are chiral polysaccharides that are abundant in nature and are employed in a variety of applications, including construction materials and food additives [146]. While they may contain different liquid crystals, lyotropic cholesteric phases are the most frequently available. Fingerprint textures with an equidistant line pattern can quickly reveal the phase's cholesteric structure and helicity. On a molecular scale, the individual polymer molecules in cellulose are chiral, and on a larger, supermolecular scale, they form helical polymers. These supermolecular structures behave in solution like a giant chiral calamitic molecule, forming the cholesteric phase, in which the supermolecular structure's long axis changes direction continually, perpendicular to the long axis, resulting into production of a cholesteric helix structure. Temperature dependency and helical pitch are straightforward for single component thermotropic cholesteric systems. Molecules with opposite configurations have opposite handedness in their pitch. It usually decreases as the temperature rises, and only a few occurrences of the opposite temperature dependence are known. The situation becomes even more difficult for thermotropic compositions, as the helicity may modify handedness as the concentration is changed. This exhibits the effects of molecule chirality and thermodynamics interaction. As previously stated, the issue becomes even more convoluted in the case of cellulose, and as a result, the temperature dependence and pitch handedness are exceedingly difficult, e.g., by applying a different achiral solvent, the cholesteric pitch of the identical cellulose molecule might have opposite handedness.

6.2.3 Cholesteric LC Phases of DNA

DNA (Deoxyribonucleic acid) and RNA (Ribonucleic acid) are prolonged biopolymers which are essential for living organisms and most viruses, along with proteins and carbohydrates. DNA is made up of nucleotides, which are composed of a nucleobase, a sugar, and a phosphate. The backbone is made up of sugar and phosphate substitution patterns. When the concentration of DNA is increased, it transforms into a liquid crystalline phase, quite similar to Blue phase [147]. The cholesteric phase and other liquid crystal phases can be seen at much higher concentrations. The helicity of the DNA double-helix and its handedness are responsible for the chirality that causes this symmetry violation. DNA from the bacteria *Escherichia coli*, for example, displays anisotropic characteristics and a polarising microscopic texture that resembles a cholesteric liquid crystalline phase.

References

[1] Sagara, Y. and T. Kato. 2009. Mechanically induced luminescence changes in molecular assemblies. *Nature Chemistry* 1: 605–610.

[2] Yu, Y., M. Nakano and T. Ikeda. 2003. Directed bending of a polymer film by light. *Nature* 425: 145.

[3] Schadt, M. 2015. Nematic liquid crystals and twisted nematic LCDs. *Liquid Crystals* 42: 646–652.

[4] Bisoyi, H.K. and L. Quan. 2016. Light-driven liquid crystalline materials: From photo-induced phase transitions and property modulations to applications. *Chemical Reviews* 116: 15089–15166.

[5] Lagerwall, J.P.F. and G.A. Scalia. 2012. A new era for liquid crystal research: Applications of liquid crystals in soft matter nano-, bio- and microtechnology. *Current Applied Physics* 12: 1387–1412.

[6] Körner, H., A. Shiota, T.J. Bunning and C.K. Ober. 1996. Orientation-on-demand thin films: Curing of liquid crystalline networks in ac electric fields. *Science* 272: 252–255.

[7] Hulvat, J.F. and S.I. Stupp. 2003. Liquid-crystal templating of conducting polymers. *Angewandte Chemie* 42: 778–781.

[8] Hentze, H.P. and E.W. Kaler. 2003. Polymerization of and within self-organized media. *Current Opinion in Colloid and Interface Science* 8: 164–178.

[9] Reinitzer, F. 1888. Zur Kenntnis des Cholesterins. *Monatsh. Chem.* 9: 421.

[10] Lehmann, O. 1889. FlieBende Kristalle. *Z. physik. Chem.* 4: 462.

[11] Demus, D. 1994. Phase types, structures and chemistry of liquid crystals. pp. 1–50. *In: Liquid Crystals.* 10.1007/978-3-662-08393-2_1.

[12] Chandrasekhar, S., B.K. Sadashiva and K.A. Suresh. 1977. Liquid crystals of disc-like molecules. *Pramana* 9: 471–480.

[13] Blinov, L.M. and V.G. Chigrinov. 1994. Liquid crystalline state. pp. 1–46. *In: Electrooptic Effects in Liquid Crystal Materials.* 10.1007/978-1-4612-2692-5_1.

[14] Tschierske, C. 1996. Molecular self-organization of amphotropic liquid crystals. *Prog. Polym. Sci.* 21: 775.

[15] Demus, D., J.W. Goodby, G.W. Gray, H.W. Spiess and V. Vill. 1998. Handbook of Liquid Crystals, Vol-3, Eds.: Wiley-VCH, New York, p. 303.

[16] Yu, L.J. and A. Saupe. 1980. Observation of a biaxial nematic phase in potassium laurate-1-decanol-water mixtures. *Phys. Rev. Lett.* 45: 1000.

[17] Tschierske, C. 2002. Liquid crystalline materials with complex mesophase morphologies. *Curr. Opin. Colloid Interface Sci.* 7: 69, 80.

[18] Friedel, G. 1922. Les états mésomorphes de la matière. *Ann. Physique.* 18: 273.

[19] Oswald, P. and P. Pieranski. 2005. Nematic and Cholesteric Liquid Crystals: Concepts and Physical Properties Illustrated by Experiments, CRC Press (google-Books-ID: 4W3YXuUFAGMC).

[20] Oswald, P. and P. Pieranski. 2005. Smectic and Columnar Liquid Crystals: Concepts and Physical Properties Illustrated by Experiments, CRC Press (google-Books-ID: f4Q1m9cLEaEC).

[21] Gordon, C.M., J.D. Holbrey, A.R. Kennedy and K.R. Seddon. 1998. Ionic liquid crystals: hexafluorophosphate salts. *J. Mater. Chem.* 8: 2627–2636.

[22] Holbrey, J.D. and K.R. Seddon. 1999. The phase behaviour of 1-alkyl-3-methylimidazolium tetrafluoroborates; ionic liquids and ionic liquid crystals. *J. Chem. Soc., Dalton Trans.* 2133–2139.

[23] Welton, T. 1999. Room-temperature ionic liquids, solvents for synthesis and catalysis. *Chem. Rev.* 99: 2071–2083.

[24] Wilkes, J.S. 2002. A short history of ionic liquids—from molten salts to neoteric solvents. *Green Chem.* 4: 73–80.

[25] Wasserscheid, P. and T. Welton (eds.). 2003. Ionic Liquids in Synthesis. Wiley-VCH: Weinheim.

[26] Hallett, J.P. and T. Welton. 2011. Room-temperature ionic liquids: solvents for synthesis and catalysis. 2. *Chem. Rev.* 111: 3508–3576.

[27] Dialkylimidazolium chloroaluminate melts—a new class of room-temperature ionic liquids for electrochemistry, spectroscopy, and synthesis. *Inorg. Chem.* 1982, 21: 1263–1264.

[28] Earle, M.J., J.M.S.S. Esperança, M.A. Gilea, J.N. Canongia Lopes, L.P.N. Rebelo, J.W. Magee, K.R. Seddon and J.A. Widegren. 2006. The distillation and volatility of ionic liquids. *Nature* 439: 831–834.

[29] Yoshizawa, M., W. Xu and C.A. Angell. 2003. Ionic liquids by proton transfer: vapor pressure, conductivity, and the relevance of ΔpKa from aqueous solutions. *J. Am. Chem. Soc.* 125: 15411–15419.

[30] Wilkes, J.S. and M.J. Zaworotko. 1992. Air and water stable 1-ethyl-3-methylimidazolium based ionic liquids. *J. Chem. Soc., Chem. Commun.* 65–967.

[31] Huddleston, J.G., A.E. Visser, W.M. Reichert, H.D. Willauer, G.A. Broker and R.D. Rogers. 2001. Characterization and comparison of hydrophilic

and hydrophobic room temperature ionic liquids incorporating the imidazolium cation. *Green Chem.* 3: 156–164.

[32] Swatloski, R.P., J.D. Holbrey and R.D. Rogers. 2003. Ionic liquids are not always green: hydrolysis of 1-butyl-3-methylimidazolium hexafluorophosphate. *Green Chem.* 5: 361–363.

[33] Cho, C.W., T.P.T. Pham, Y.C. Jeon and Y.S. Yun. 2008. Influence of anions on the toxic effects of ionic liquids to a phytoplankton selenastrum capricornutum. *Green Chem.* 10: 67–72.

[34] Freire, M.G., C.M. Neves, I.M. Marrucho, J.A.P. Coutinho and A.M. Fernandes. 2010. Hydrolysis of tetrafluoroborate and hexafluorophosphate counter ions in imidazolium-based ionic liquids. *J. Phys. Chem. A* 114: 3744–3749.

[35] Ohno, H. 2005. Electrochemical Aspects of Ionic Liquids. Ed.; Wiley-VCH: New York.

[36] Nakagawa, H., S. Izuchi, K. Kuwana, T. Nukuda and Y. Aihara. 2003. Liquid and polymer gel electrolytes for lithium batteries composed of room-temperature molten salt doped by lithium salt. *J. Electrochem. Soc.* 150: A695–A700.

[37] Endres, F. and S.Z. El Abedin. 2006. Air and water stable ionic liquids in physical chemistry. *Phys. Chem. Chem. Phys.* 8: 2101–2116.

[38] Papageorgiou, N., Y. Athanassov, M. Armand, P. Bonhôte, H. Pettersson, A. Azam and M. Grätzel. 1996. The performance and stability of ambient temperature molten salts for solar cell applications. *J. Electrochem. Soc.* 143: 3099–3108.

[39] Wang, P., S.M. Zakeeruddin, J.E. Moser and M. Grätzel. 2003. A new ionic liquid electrolyte enhances the conversion efficiency of dye-sensitized solar cells. *J. Phys. Chem. B* 107: 13280–13285.

[40] Marisa, C. Buzzeo, Russell G. Evans and Richard G. Compton. Prof. 2004. Non-haloaluminate room-temperature ionic liquids in electrochemistry—A review. *Chem. Phys. Chem.* 5: 1106–1120.

[41] Zhao, Y.G. and T.J. Van der Noot. 1997. Electrodeposition of aluminium from nonaqueous organic electrolytic systems and room temperature molten salts. *Electrochim. Acta* 42: 3–13.

[42] Endres, F. 2002. Ionic liquids: solvents for the electrodeposition of metals and semiconductors. *ChemPhysChem.* 3: 144–154.

[43] Endres, F., M. Bukowski, R. Hempelmann and H. Natter. 2003. Electrodeposition of nanocrystalline metals and alloys from ionic liquids. *Angew. Chem., Int. Ed.* 42: 3428–3430.

[44] Endres, F. 2004. Ionic liquids: promising solvents for electrochemistry. *Z. Phys. Chem.* 218: 255–283.

[45] Brooks, N.R., S. Schaltin, K. Van Hecke, L. Van Meervelt, K. Binnemans and J. Fransaer. 2011. Copper(I)-containing ionic liquids for high-rate electrodeposition. *Chem. - Eur. J.* 17: 5054–5059.

[46] Nockemann, P., E. Beurer, K. Driesen, R. Van Deun, K. Van Hecke, L. Van Meervelt and K. Binnemans. 2005. Photostability of a highly luminescent europium β-diketonate complex in imidazolium ionic liquids. *Chem. Commun.* 4354–4356.

[47] Binnemans, K. 2007. Lanthanides and actinides in ionic liquids. *Chem. Rev.* 107: 2592–2614.

[48] Antonietti, M., D.B. Kuang, B. Smarsly and Y. Zhou. 2004. Ionic liquids for the convenient synthesis of functional nanoparticles and other inorganic nanostructures. *Angew. Chem., Int. Ed.* 43: 4988–4992.

[49] Zhou, Y. 2005. Recent advances in ionic liquids for synthesis of inorganic nanomaterials. *Curr. Nanosci.* 1: 35–42.

[50] Taubert, A. 2005. Inorganic materials synthesis—a bright future for ionic liquids? *Acta Chim. Slov.* 52: 183–186.

[51] Reichert, W.M., J.D. Holbrey, K.B. Vigour, T.D. Morgan, G.A. Broker and R.D. Rogers. 2006. Approaches to crystallization from ionic liquids: complex solvents-complex results, or, a strategy for controlled formation of new supramolecular architectures? *Chem. Commun.* 4767–4779.

[52] Taubert, A. and Z. Li. 2007. Inorganic materials from ionic liquids. *Dalton Trans.* 723–727.

[53] Plechkova, N.V. and K.R. Seddon. 2008. Applications of ionic liquids in the chemical industry. *Chem. Soc. Rev.* 37: 123–150.

[54] Cooper, E.R., C.D. Andrews, P.S. Wheatley, P.B. Webb, P. Wormald and R.E. Morris. 2004. Ionic liquids and eutectic mixtures as solvent and template in synthesis of zeolite analogues. *Nature* 430: 1012–1016.

[55] Parnham, E.R. and R.E. Morris. 2006. 1-alkyl-3-methyl imidazolium bromide ionic liquids in the ionothermal synthesis of aluminium phosphate molecular sieves. *Chem. Mater.* 18: 4882–4887.

[56] Parnham, E.R. and R.E. Morris. 2007. Ionothermal synthesis of zeolites, metal-organic frameworks, and inorganic-organic hybrids. *Acc. Chem. Res.* 40: 1005–1013.

[57] Morris, R.E. 2009. Ionothermal synthesis-ionic liquids as functional solvents in the preparation of crystalline materials. *Chem. Commun.* 2990–2998.

[58] Morris, R.E. and X. Bu. 2010. Induction of chiral porous solids containing only achiral building blocks. *Nat. Chem.* 2: 353–361.

[59] Kore, R. and R. Srivastava. 2012. Synthesis of zeolite beta, MFI, and MTW using imidazole, piperidine, and pyridine based quaternary ammonium salts as structure directing agents. *RSC Adv.* 2: 10072–10084.

[60] Chen, Z., T.L. Greaves, R.A. Caruso and C.J. Drummond. 2012. Long-range ordered lyotropic liquid crystals in intermediate-range ordered protic ionic liquid used as templates for hierarchically porous silica. *J. Mater. Chem.* 22: 10069–10076.

[61] Yoshio, M., T. Mukai, K. Kanie, M. Yoshizawa, H. Ohno and T. Kato. 2002. Layered ionic liquids: anisotropic ion conduction in new self-organized liquid-crystalline materials. *Adv. Mater.* 14: 351–354.

[62] Yoshio, M., T. Kato, T. Mukai, M. Yoshizawa and H. Ohno. 2004. Self-assembly of an ionic liquid and a hydroxyl-terminated liquid crystal: anisotropic ion conduction in layered nanostructures. *Mol. Cryst. Liq. Cryst.* 413: 2235–2244.

[63] Shimura, H., M. Yoshio, K. Hoshino, T. Mukai, H. Ohno and T. Kato. 2008. Noncovalent approach to one-dimensional ion conductors: enhancement of ionic conductivities in nanostructured columnar liquid crystals. *J. Am. Chem. Soc.* 130: 1759–1765.

[64] Greaves, T.L. and C.J. Drummond. 2008. Ionic liquids as amphiphile self-assembly media. *Chem. Soc. Rev.* 37: 1709–1726.

[65] Greaves, T.L., A. Weerawardena, I. Krodkiewska and C.J. Drummond. 2008. Protic ionic liquids: physicochemical properties and behavior as amphiphile self-assembly solvents. *J. Phys. Chem. B* 112: 896–905.

[66] Mulet, X., D.F. Kennedy, T.L. Greaves, L.J. Waddington, A. Hawley, N. Kirby and C.J. Drummond. 2010. Diverse ordered 3d nanostructured amphiphile self-assembly materials found in protic ionic liquids. *J. Phys. Chem. Lett.* 1: 2651–2654.

[67] Ichikawa, T., M. Yoshio, S. Taguchi, J. Kagimoto, H. Ohno and T. Kato. 2012. Co-organisation of ionic liquids with amphiphilic diethanolamines: construction of 3D continuous ionic nanochannels through the induction of liquid-crystalline bicontinuous cubic phases. *Chem. Sci.* 3: 2001–2008.

[68] Chen, Z., T.L. Greaves, C. Fong, R.A. Caruso and C.J. Drummond. 2012. Lyotropic liquid crystalline phase behaviour in amphiphile-protic ionic liquid systems. *Phys. Chem. Chem. Phys.* 14: 3825–3836.

[69] Ichikawa, T., K. Fujimura, M. Yoshio, T. Kato and H. Ohno. 2013. Designer lyotropic liquid-crystalline systems containing amino acid ionic liquids as self-organisation media of amphiphiles. *Chem. Commun.* 49: 11746–11748.

[70] Yamashita, A., M. Yoshio, S. Shimizu, T. Ichikawa, H. Ohno and T. Kato. 2015. Columnar nanostructured polymer films containing ionic liquids in supramolecular one-dimensional nanochannels. *J. Polym. Sci., Part A: Polym. Chem.* 53: 366–371.

[71] Greaves, T.L. and C.J. Drummond. 2013. Solvent nanostructure, the solvophobic effect and amphiphile self-assembly in ionic liquids. *Chem. Soc. Rev.* 42: 1096–1120.

[72] Greaves, T.L. and C.J. Drummond. 2015. Protic ionic liquids: evolving structure-property relationships and expanding applications. *Chem. Rev.* 115: 11379–11448.

[73] Del Pópolo, M.G. and G.A. Voth. 2004. On the structure and dynamics of ionic liquids. *J. Phys. Chem. B* 108: 1744–1752.

[74] Urahata, S.M. and M.C.C. Ribeiro. 2004. Structure of ionic liquids of 1-alkyl-3-ethylimidazolium cations: a systematic computer simulation study. *J. Chem. Phys.* 120: 1855–1863.

[75] Tokuda, H., K. Hayamizu, K. Ishii, M.A.B.H. Susan and M. Watanabe. 2005. Physicochemical properties and structures of room temperature ionic

liquids. 2. Variation of alkyl chain length in imidazolium cation. *J. Phys. Chem. B* 109: 6103–6110.

[76] Deetlefs, M., C. Hardacre, M. Nieuwenhuyzen, A.A.H. Pádua, O. Sheppard and A.K. Soper. 2006. Liquid structure of the ionic liquid 1,3-dimethylimidazolium bis{(trifluoromethyl)sulfonyl}amide. *J. Phys. Chem. B* 110: 12055–12061.

[77] Canongia Lopes, J.N., M.F.C. Gomes and A.A.H. Pádua. 2006. Nonpolar, polar, and associating solutes in ionic liquids. *J. Phys. Chem. B* 110: 16816–16818.

[78] Triolo, A., A. Mandanici, O. Russina, V. Rodriguez-Mora, M. Cutroni, C. Hardacre, M. Nieuwenhuyzen, H.-J. Bleif, L. Keller and M.A. Ramos. 2006. Thermodynamics, structure, and dynamics in room temperature ionic liquids: the case of 1-butyl-3-methyl imidazolium hexafluorophosphate ([Bmim][PF6]). *J. Phys. Chem. B* 110: 21357–21364.

[79] Hyun, B.R., S.V. Dzyuba, R.A. Bartsch and E.L. Quitevis. 2002. Intermolecular dynamics of room-temperature ionic liquids: femtosecond optical kerr effect measurements on 1-alkyl-3-methylimidazolium bis((trifluoromethyl)sulfonyl)imides. *J. Phys. Chem. A* 106: 7579–7585.

[80] Xiao, D., J.R. Rajian, S. Li, R.A. Bartsch and E.L. Quitevis. 2006. Additivity in the optical kerr effect spectra of binary ionic liquid mixtures: implications for nanostructural organization. *J. Phys. Chem. B* 110: 16174–16178.

[81] Xiao, D., J.R. Rajian, A. Cady, S. Li, R.A. Bartsch and E.L. Quitevis. 2007. Nanostructural organization and anion effects on the temperature dependence of the optical kerr effect spectra of ionic liquids. *J. Phys. Chem. B* 111: 4669–4677.

[82] Triolo, A., O. Russina, H.-J. Bleif and E. Di Cola. 2007. Nanoscale segregation in room temperature ionic liquids. *J. Phys. Chem. B* 111: 4641–4644.

[83] Wang, Y., W. Jiang and G.A. Voth. 2007. Spatial heterogeneity in ionic liquids. pp. 272–307. *In*: Brennecke, J.F., R.D. Rogers and K.R. Seddon (eds.). Ionic Liquids IV: Not Just Solvents Anymore. ACS Symposium Series; American Chemical Society: WA, Vol. 975.

[84] Pádua, A.A.H., M.F. Gomes and J.N. Canongia Lopes. 2007. Molecular solutes in ionic liquids: a structural perspective. *Acc. Chem. Res.* 40: 1087–1096.

[85] Annapureddy, H.V., H.K. Kashyap, P.M. De Biase and C.J. Margulis. 2010. What is the origin of the prepeak in the x-ray scattering of imidazolium-based room-temperature ionic liquids? *J. Phys. Chem. B* 114: 16838–16846.

[86] Wang, Y.T. and G.A. Voth. 2005. Unique spatial heterogeneity in ionic liquids. *J. Am. Chem. Soc.* 127: 12192–12193.

[87] Wang, Y.T. and G.A. Voth. 2006. Tail aggregation and domain diffusion in ionic liquids. *J. Phys. Chem. B* 110: 18601–18608.

[88] Triolo, A., O. Russina, H.-J. Bleif and E. Di Cola. 2007. Nanoscale segregation in room temperature ionic liquids. *J. Phys. Chem. B* 111: 4641–4644.

[89] Rebelo, L.P.N., J.N. Canongia Lopes, J.M.S.S. Esperança, H.J.R. Guedes, J. Lachwa, V. Najdanovic-Visak and Z.P. Visak. 2007. Accounting for the Unique, doubly dual nature of ionic liquids from a molecular thermodynamic, and modeling standpoint. *Acc. Chem. Res.* 40: 1114–1121.

[90] Greaves, T.L., D.F. Kennedy, S.T. Mudie and C.J. Drummond. 2010. Diversity observed in the nanostructure of protic ionic liquids. *J. Phys. Chem. B* 114: 10022–10031.

[91] Burrell, G.L., N.F. Dunlop and F. Separovic. 2010. Non-newtonian viscous shear thinning in ionic liquids. *Soft Matter* 6: 2080–2086.

[92] Zhao, H., R. Shi and Y. Wang. 2011. Nanoscale tail aggregation in ionic liquids: roles of electrostatic and van der waals interactions. *Commun. Theor. Phys.* 56: 499–503.

[93] Fujii, K., R. Kanzaki, T. Takamuku, Y. Kameda, S. Kohara, M. Kanakubo, M. Shibayama, S. Ishiguro and Y. Umebayashi. 2011. Experimental evidences for molecular origin of low-q peak in neutron/X-ray scattering of 1-alkyl-3-methylimidazolium bis-(trifluoromethanesulfonyl)amide ionic liquids. *J. Chem. Phys.* 135: 244502.

[94] Hayes, R., S. Imberti, G.G. Warr and R. Atkin. 2011. Pronounced sponge-like nanostructure in propylammonium nitrate. Phys. *Chem. Chem. Phys.* 13: 13544–13551.

[95] Yago, T. and M. Wakasa. 2011. Nanoscale structure of ionic liquid and diffusion process as studied by the MFE probe. *J. Phys. Chem. C* 115: 2673–2678.

[96] Dupont, J. 2011. From molten salts to ionic liquids: a "nano" journey. *Acc. Chem. Res.* 44: 1223–1231.

[97] Rocha, M.A.A., C.F.R.A.C. Lima, L.R. Gomes, B. Schröder, J.A.P. Coutinho, I.M. Marrucho, J.M.S.S. Esperança, L.P.N. Rebelo, K. Shimizu, J.N. Canongia Lopes et al. 2011. High-accuracy vapor pressure data of the extended [C_nC1im][NTf2] ionic liquid series: trend changes and structural shifts. *J. Phys. Chem. B* 115: 10919–10926.

[98] Russina, O., A. Triolo, L. Gontrani and R. Caminiti. 2012. Mesoscopic structural heterogeneities in room-temperature ionic liquids. *J. Phys. Chem. Lett.* 3: 27–33.

[99] Hardacre, C., J.D. Holbrey, C.L. Mullan, T.G. Youngs and D.T. Bowron. 2010. Small angle neutron scattering from 1-alkyl-3-methylimidazolium hexafluorophosphate ionic liquids ([Cnmim]- [PF_6], n = 4, 6, and 8). *J. Chem. Phys.* 133: 074510.

[100] Ji, Y., R. Shi, Y. Wang and G. Saielli. 2013. Effect of the chain length on the structure of ionic liquids: from spatial heterogeneity to ionic liquid crystals. *J. Phys. Chem. B* 117: 1104–1109.

[101] Bradley, A.E., C. Hardacre, J.D. Holbrey, S. Johnston, S.E.J. McMath and M. Nieuwenhuyzen. 2002. Small-angle X-ray scattering studies of liquid crystalline 1-alkyl-3-methylimidazolium salts. *Chem. Mater.* 14: 629–635.

[102] De Roche, J., C.M. Gordon, C.T. Imrie, M.D. Ingram, A.R. Kennedy, F. Lo Celso and A. Triolo. 2003. Application of complementary experimental

techniques to characterization of the phase behavior of [C$_{16}$mim][PF$_6$] and [C$_{14}$mim][PF$_6$]. *Chem. Mater.* 15: 3089–3097.

[103] Abdallah, D.J., A. Robertson, H.F. Hsu and R.G. Weiss. 2000. Smectic liquid-crystalline phases of quaternary group VA (Especially Phosphonium) salts with three equivalent long n-alkyl chains. how do layered assemblies form in liquid-crystalline and crystalline phases? *J. Am. Chem. Soc.* 122: 3053–3062.

[104] Tripathi, C.S.P., J. Leys, P. Losada-Pérez, K. Lava, K. Binnemans, C. Glorieux and J. Thoen. 2013. Adiabatic scanning calorimetry study of ionic liquid crystals with highly ordered crystal smectic phases. *Liq. Cryst.* 40: 329–338.

[105] Yan, T., Y. Wang and C. Knox. 2010. On the structure of ionic liquids: comparisons between electronically polarizable and nonpolarizable models I. *J. Phys. Chem. B* 114: 6905–6921.

[106] Yan, T., Y. Wang and C. Knox. 2010. On the dynamics of ionic liquids: comparisons between electronically polarizable and nonpolarizable models II. *J. Phys. Chem. B* 114: 6886–6904.

[107] Smith, G.D., O. Borodin, L. Li, H. Kim, Q. Liu, J.E. Bara, D.L. Gin and R.D. Noble. 2008. A comparison of ether- and alkyl-derivatized imidazolium-based room-temperature ionic liquids: a molecular dynamics simulation study. *Phys. Chem. Chem. Phys.* 10: 6301–6312.

[108] Shirota, H., H. Fukazawa, T. Fujisawa and J.F. Wishart. 2010. Heavy atom substitution effects in non-aromatic ionic liquids: ultrafast dynamics and physical properties. *J. Phys. Chem. B* 114: 9400–9412.

[109] Siqueira, L.J. and M.C. Ribeiro. 2011. Charge ordering and intermediate range order in ammonium ionic liquids. *J. Chem. Phys.* 135: 204506.

[110] Binnemans, K. 2005. Ionic liquid crystals. *Chem. Rev.* 105: 4148–4204.

[111] Axenov, K.V. and S. Laschat. 2011. Thermotropic ionic liquid crystals. *Materials* 4: 206–259.

[112] Mansueto, M. and S. Laschat. 2014. Ionic liquid crystals. pp. 231–280. *In*: Goodby, J.W., P.J. Collings, T. Kato, C. Tschierske, H. Gleeson and P. Raynes. (eds.). Handbook of Liquid Crystals. Vol. 6: Nanostructured and Amphiphilic Liquid Crystals, 2nd ed. Wiley-VCH: Weinheim.

[113] Tschierske, C. 1996. Molecular self-organization of amphotropic liquid crystals. *Prog. Polym. Sci.* 21: 775–852.

[114] Neve, F. 1996. Transition metal based ionic mesogens. *Adv. Mater.* 8: 277–289.

[115] Skoulios, A. and D. Guillon. 1988. Amphiphilic character and liquid crystallinity. *Mol. Cryst. Liq. Cryst.* 165: 317–332.

[116] Paleos, C.M. 1994. Thermotropic liquid crystals derived from amphiphilic mesogens. *Mol. Cryst. Liq. Cryst. Sci. Technol., Sect. A* 243: 159–183.

[117] Somashekar, R. 1987. Mesomorphic behavior of n-4-hexadecylpyridinium chloride. *Mol. Cryst. Liq. Cryst.* 146: 225–233.

[118] Knight, G.A. and B.D. Shaw. 1938. Long-chain alkylpyridines and their derivatives. New examples of liquid crystals. *J. Chem. Soc.* 682–683.

[119] Hardacre, C. 2002. Order in the liquid state and structure. pp. 127–152. *In*: Wasserscheid, P. and T. Welton (eds.). Ionic Liquids in Synthesis. Wiley-VCH: Weinheim.

[120] Kato, T. and M. Yoshio. 2005. Liquid crystalline ionic liquids. pp 307–320. *In*: Ohno, H. (ed.). Electrochemical Aspects of Ionic Liquids. Wiley-VCH: New York.

[121] Lin, I.J.B. and C.S. Vasam. 2005. Metal-containing ionic liquids and ionic liquid crystals based on imidazolium moiety. *J. Organomet. Chem.* 690: 3498–3512.

[122] Kato, T. and M. Yoshio. 2011. Liquid crystalline ionic liquids. pp. 375–392. *In*: Ohno, H. (ed.). Electrochemical Aspects of Ionic Liquids, 2nd ed. Wiley-VCH: Hoboken, NJ.

[123] Blunk, D., K. Praefcke and V. Vill. 1998. Amphotropic liquid crystals. pp. 305–340. *In*: Demus, D., J.W. Goodby, G.W. Gray, H.-W. Spiess and V. Vill (eds.). Handbook of Liquid Crystals. Vol. 3: High Molecular Weight Liquid Crystals. Wiley-VCH: Weinheim.

[124] Fairhurst, C., S. Fuller, J. Gray, M.C. Holmes and G.J.T. Tiddy. 1998. Lyotropic surfactant liquid crystals. pp. 341–392. *In*: Demus, D., J.W. Goodby, G.W. Gray, H.-W. Spiess and V. Vill (eds.). Handbook of Liquid Crystals. Vol. 3: High Molecular Weight Liquid Crystals. Wiley-VCH: Weinheim.

[125] Imrie, C.T. and G.R. Luckhurst. 1998. Liquid crystal dimers and oligomers. pp. 801–834. *In*: Demus, D., J.W. Goodby, G.W. Gray, H.-W. Spiess and V. Vill (eds.). Handbook of Liquid Crystals. Vol. 2B: Low Molecular Weight Liquid Crystals II. Wiley-VCH: Weinheim.

[126] Imrie, C.T. 1999. Liquid crystal dimers. pp. 149–191. *In*: Mingos, D.M.P. (ed.). Liquid Crystals II. Springer: Berlin, Heidelberg.

[127] Imrie, C.T. and P.A. Henderson. 2007. Liquid crystal dimers and higher oligomers: between monomers and polymers. *Chem. Soc. Rev.* 36: 2096–2124.

[128] Tschierske, C. 1998. Non-conventional liquid crystals—the importance of micro-segregation for self-organisation. *J. Mater. Chem.* 8: 1485–1508.

[129] Faul, C.F.J. and M. Antonietti. 2002. Facile synthesis of optically functional, highly organized nanostructures: dye-surfactant complexes. *Chem. - Eur. J.* 8: 2764–2768.

[130] Faul, C.F.J. and M. Antonietti. 2003. Ionic self-assembly: facile synthesis of supramolecular materials. *Adv. Mater.* 15: 673–683.

[131] Faul, C.F.J. 2006. Liquid-crystalline materials by the ionic self-assembly route. *Mol. Cryst. Liq. Cryst.* 450: 255–265.

[132] Zhang, T., J. Brown, R.J. Oakley and C.F.J. Faul. 2009. Towards functional nanostructures: ionic self-assembly of polyoxometalates and surfactants. *Curr. Opin. Colloid Interface Sci.* 14: 62–70.

[133] Camerel, F. and C.F.J. Faul. 2003. Combination of ionic self-assembly and hydrogen bonding as a tool for the synthesis of liquid-crystalline materials and organogelators from a simple building block. *Chem. Commun.* 1958–1959.

[134] Hosseini, M.W. 2005. Molecular tectonics: from simple tectons to complex molecular networks. *Acc. Chem. Res.* 38: 313–323.

[135] Kitzerow, H.-S. 2001. Chirality in Liquid Crystals; Bahr, C. Ed.; Springer Verlag: New York, NY, USA.

[136] Dierking, I. 2003. Textures of Liquid Crystals. Wiley-VCH: Weinheim, Germany.

[137] Keating, P.N. 1969. A theory of the cholesteric mesophase. *Mol. Cryst.* 8: 315–326.

[138] De Gennes, P.G. 1973. Some remarks on the polymorphism of smectics. *Mol. Cryst. Liq. Cryst.* 21: 49–76.

[139] Chen, J.H. and T.C. Lubensky. 1976. Landau-ginzburg mean field theory for the nematic to smectic-C and nematic to smectic-A phase transition. *Phys. Rev. A* 14: 1202–1207.

[140] Chu, K.C. and W.L. McMillan. 1977. Unified landau theory for the nematic, smectic A, and smectic C phases of liquid crystals. *Phys. Rev. A* 15: 1181–1187.

[141] Yoon, H.G., I. Dierking and H.F. Gleeson. 2010. Cholesteric pitch divergence near smectic phase transitions. *Phys. Rev. E* 82: 011705:1–011705:9.

[142] Dierking, I., F. Giesselmann, P. Zugenmaier, W. Kuczynski, S.T. Lagerwall and B. Stebler. 1993. Investigations of the structure of a cholesteric phase with a temperature induced helix inversion and of the succeeding SC* phase in thin liquid crystal cells. *Liq. Cryst.* 13: 45–55.

[143] Styring, P., J.D. Vuijk, I. Nishiyama, A.J. Slaney and J.W. Goodby. 1993. Inversion of chirality dependent properties in optically active liquid crystals. *J. Mater. Chem.* 3: 399–405.

[144] Finkelmann, H. and H. Stegemeyer. 1978. Temperature dependence of the intrinsic pitch in induced cholesteric systems. *Ber. Bunsenges. Phys. Chem.* 82: 1302–1308.

[145] Dierking, I., F. Giesselmann, P. Zugenmaier, K. Mohr, H. Zaschke, W. Kuczynski. 1995. New diastereomeric compound with cholesteric twist inversion. *Liq. Cryst.* 18: 443–449.

[146] Zugenmaier, P. 2014. Cellulosic liquid crystals. *In*: Goodby, J.W., P.J. Collings, T. Kato, C. Tschierske, H.F. Gleeson and P. Raynes (eds.). Handbook of Liquid Crystals. Wiley-VCH: Weinheim, Germany, Volume 7.

[147] Zanchetta, G., M. Nakata, M. Buscaglia, T. Bellini and N.A. Clark. 2008. Phase separation and liquid crystallization of complementary sequences in mixtures of nanoDNA oligomers. *Proc. Natl. Acad. Sci. USA* 105: 1111–1117.

CHAPTER 2

Micro- and Nano-particles Doped Liquid Crystals

1. Introduction

Liquid crystals are often identified with the development of the flat panel television and computer screens that we all use on a daily basis. Despite their enormous success in this area, liquid crystal research is by far not exhausted and has reinvented itself, spearheading into other fields of research, due to their properties of self-organization, their fascinating optic and electro-optic properties, and their easy deformability and reorientation via electric, magnetic, mechanical and other external fields. Liquid crystals are also being employed as a vehicle to study fundamental physical questions, and proceeding into the areas of biology, nature and life. In this Special Issue of Nanomaterials, illustrative examples are introduced, which draw on aspects of self-organization of liquid crystals, colloidal ordering of nanoparticles, and the formation of anisotropic, liquid crystalline phases from nanoparticles. Liquid crystals [1–3] are partially ordered, anisotropic fluids, which are thermodynamically located between the three-dimensional solid crystal and the flow governed liquid. They exhibit orientational or low dimensional positional order of their long molecular axis or the molecular centres of mass, respectively, which results in anisotropic physical properties, such as refractive index, viscosity, elastic constant, electric conductivity, or magnetic susceptibility, while retaining the ability to flow. Liquid crystals are part of the ever growing and increasingly important family of soft condensed matter materials [4–8]. Thermotropic dispersions and lyotropic liquid crystalline behaviour has also been reported for carbon based materials, for example involving single-walled and multi-walled nanotubes [9–14] for electrically and magnetically addressable molecular switches. Also lyotropic graphene oxide [15–18] has been explored in a variety of host liquids for possible electro-optic applications based on the Kerr effect. Further reports discuss inorganic nanorods [19–21], ferroelectric particles [22] and magnetic nanorods [23] and platelets for ferromagnetic nematics. Also, the incorporation of gold nanoparticles [24, 25] into liquid crystals or indeed mesogenic molecules has become popular, especially for applications in plasmonics. Furthermore, carbon materials such as fullerenes [26] are also incorporated into liquid crystal forming molecules.

The simplest way to classify nanomaterials used in combination with liquid crystal materials or the liquid crystalline state is by using their shape. Three "shape families" of nanomaterials have emerged as the most popular, and sorted from the highest to the lowest frequency of appearance in published studies; these are zero-dimensional (quasi-spherical) nanoparticles, one-dimensional (rod or wirelike) nanomaterials such as nanorods, nanotubes, or nanowires, and two-dimensional (disc-like) nanomaterials such as nanosheets, nanoplatelets, or nanodiscs. Scheme 2.1 presents the different types of nano-particles in LCs with wider aspect of shapes and chemical compositions. The aforementioned frequency of the use of these nanomaterial shapes is best attributed to two factors: (a) the ease with which these nanoparticle shapes can be synthesized in the laboratory and (b) the availability of these nanomaterials from commercial sources. It cannot be the aim of this review to cover all of the different nanomaterials used so far, but some of the most commonly investigated ones will be introduced in more detail. For zero-dimensional nanoparticles, emphasis will be put on metallic nanoparticles (mainly gold), semiconductor quantum dots, as well as magnetic (different iron oxides) and ferroelectric nanoparticles. In the area of one dimensional nanomaterials, metal and semiconductor nanorods and nanowires as well as carbon nanotubes will be briefly discussed, and for

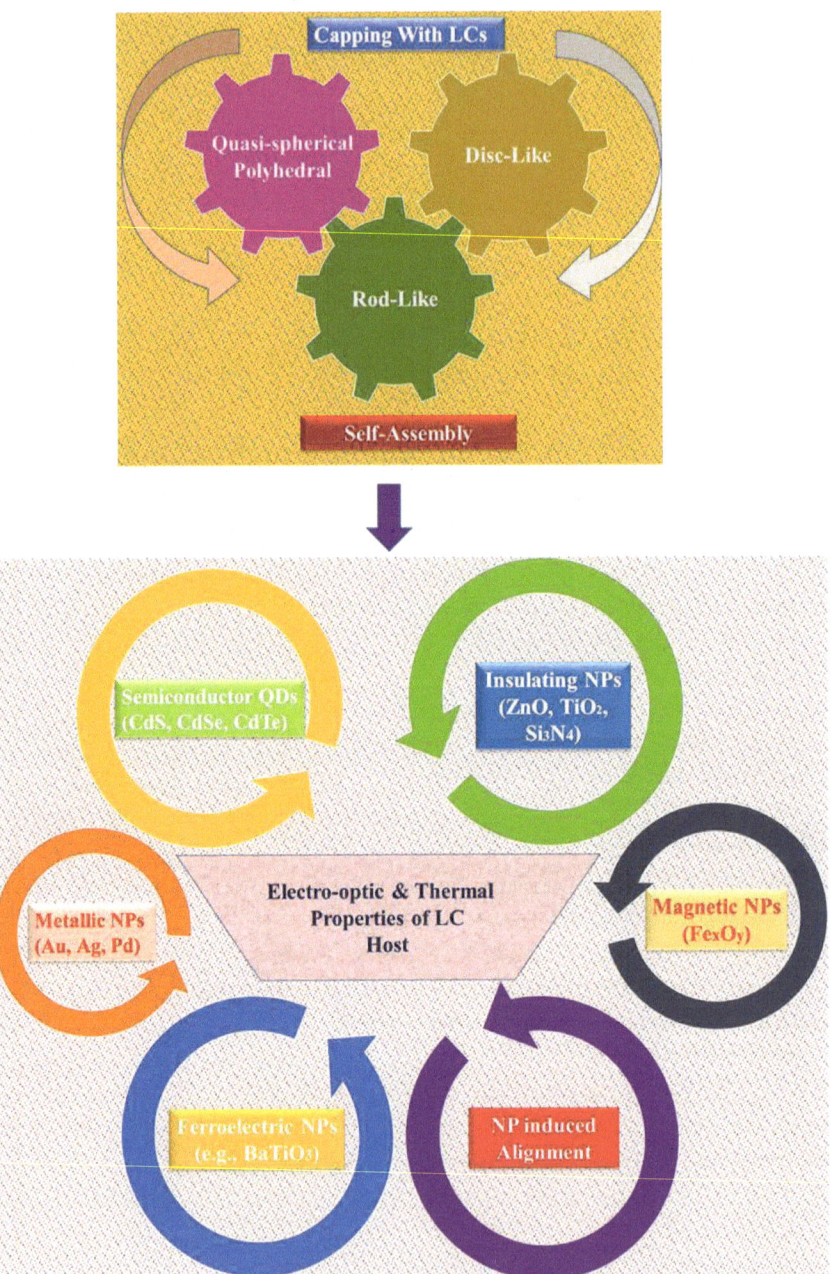

Scheme 2.1. Graphical abstract of nano-particles in liquid crystal.

two-dimensional nanomaterials only nanoclay. Finally, researchers active in the field are advised to seek further information about these and other nanomaterials in the following, very insightful review articles [27–33]. This chapter will focus on the important synthesis and characterizations methods of various important nanomaterials of liquid crystal study. The later part will discuss on the potential applications of these nanomaterials as additives or dopants in liquid crystalline systems.

2. Zero-dimensional Nanoparticles

2.1 Metal Nanoparticles

Most of the work dealing either with the assembly of metal nanoparticles or with metal nanoparticles as additives in liquid crystals has been carried out using nanoparticles with gold cores. Such gold nanoparticles display unusual and unique size-dependent chemical and physical properties; their surface can be passivated in situ or via place exchange reactions with molecules ranging from simple alkanethiols, phosphines, and amines to more complex molecular structures including dendrimers, polymers, and peptides. Less attention has so far been paid to other metals such as silver, copper, palladium, or platinum. However, the preparation of some of the latter is possible in principle, and often done, using similar methods as described here for nanoparticles with gold cores. Gold nanoparticles can be produced by a number of different methods including sonochemical syntheses [34], syntheses using ionic liquids [35–38], and many others. Each method claims a certain advantage in terms of being a green or surfactant-free method, or a method with good size-control, etc., but the majority of research groups to date use one of the two revolutionary methods developed by Brust and Schiffrin in the mid-1990s. The so-called one- [39] and two-phase Brust–Schiffrin reactions [40] providing thiol-capped gold nanoparticles or monolayer-protected clusters allow for a reasonable control over size and size-distribution, and can be carried out using a great variety of functionalized thiols [41, 42], given that they survive the reductive power of sodium borohydride

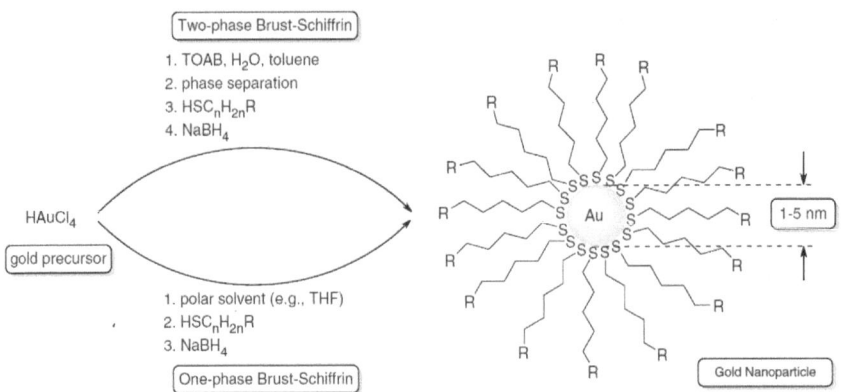

R = –CH$_3$, –C$_m$H$_{2m+1}$, functional group, functional molecule (e.g., chiral, mesogenic, pro-mesogenic)
TOAB = tetraoctylammonium bromide

via:

(III)   (I)

[NR$_4$][AuCl$_4$] + 2 R'SH ⟶ [NR$_4$][AuCl$_2$] + R'SSR' + 2 HCl

Scheme 2.2. The one- and two-phase Brust–Schiffrin methods for the synthesis of thiol-capped gold nanoparticles, as well as the initial stage of the reaction before the reduction step with NaBH$_4$ [43].

(NaBH$_4$) in the final step of the reaction. The mechanisms of these reactions have been debated in literature for many years, and recent work by Lennox et al. using 1H NMR spectroscopy now shows that the Au(I) thiolate polymer intermediate, often presumed, is not formed prior to the reductive step with NaBH$_4$ in both the one- and two-phase Brust–Schiffrin method (see Scheme 2.2) [43].

Other commonly used methods for obtaining functionalized gold nanoparticles include (a) the so-called place exchange reaction (often thiol-for-thiol exchange) on the nanoparticle surface [44, 45], (b) post-synthetic modifications (e.g., using a reactive R group on the nanoparticle corona as in Scheme 2.2), and (c) the displacement of weaker ligands or weaker surface protective agents with functional, stronger binding ligands (e.g., a thiol-for-dimethyl aminopyridine [46, 47] or thiol-for-triphenylphosphine exchange to obtain gold nanoparticles with low polydispersity indices [48]). Unfortunately, challenges such as repeatability, chemical purity, and homogeneity of nanoparticle samples are commonly disregarded by groups working in the sector. Even when using generally acknowledged, published processes, batches of nanoparticles made in different laboratories or at different times are not always equal in terms of purity, size, or size dispersion. Each newly prepared batch of nanoparticles should be required to characterise all chemical and physical properties, as well as report all data necessary to prove unequivocal purity and size/size distribution, including, but not limited to, 1H NMR (absence of free, non-bound ligands, ammonium salts, or other impurities/reagents) and elemental analysis and/or inductively coupled plasma spectroscopy, ICP-OES/MS. In this regard, TEM, X-ray diffraction/scattering, or DLS offer information on purity and monolayer coverage in combination with size information provided by TEM, X-ray diffraction/scattering, or DLS. The chemical and thermal (in)stability of metal nanoparticles, as well as the suggested mobility of surface bound capping agents, add to the complexity of the situation (usually thiols). A number of groups [49] have investigated the chemical stability or instability of thiolcapped metal nanoparticles towards oxidation (i.e., oxidation of surface-bound thiols in air or in the presence of other oxidants) [50], halides [51], and alkaline metal ions using TEM, UV-vis, NMR, and X-ray photoelectron spectroscopy (XPS) [52], and this work highlights the importance of evaluating purity of nanoparticles.

2.2 Quantum Dots

Another type of zero-dimensional nanoparticle is quantum dots. Due to the various possible elemental compositions such as CdS, CdSe, CdTe, ZnS, ZnSe, PbS, and many others, including core-shell [53–59] and alloy-type quantum dots [60] including elements such as In, Hg [61], As [62], and several lanthanides [63], the variety of quantum dots based on II–VI, III–V, and IV–VI semiconductor materials is far greater than that for metal nanoparticles. Many hydrophilic and hydrophobic quantum dots are now commercially available as suspensions in organic or aqueous solvents, but researchers can choose between organometallic syntheses based on high-temperature thermolysis [64] of a precursor or sol-gel type aqueous phase syntheses [65] for a specific coating motif or quantum dot size.

CdS, CdSe, and CdTe capped with trioctylphosphine/ trioctylphosphine oxide (TOP/TOPO) [66, 67], CdS and CdTe stabilised with thiols [68], and CdSe protected with amines [69] are all examples of quantum dots utilised in liquid crystal systems (with sizes ranging from a few to tens of nanometers). These quantum dots can be generated or size divided into batches encompassing practically the whole visible spectral range from 400 to 750 nm with, in part, high photoluminescence quantum efficiencies, depending on the type of synthesis (some stable in air [70], others not [71]). Weller et al. demonstrated a highly effective synthesis for hydrophilic, thiol-capped CdTe quantum dots [72, 73], which may be converted to lipophilic, alkanethiol-stabilized CdTe quantum dots by a place exchange process similar to that described above for metal nanoparticles [74]. Hydrophobic or otherwise functionalized CdS [75] or CdSe quantum dots [76] have also been successfully produced using a similar technique. The stability of these quantum dots in liquid crystal applications is a serious concern. Weller et al. demonstrated that thiol-stabilized cadmium chalcogenide quantum dots are unstable at low particle concentrations (10 mM) utilising NMR spectroscopy, UV-vis spectrophotometry, and analytical ultracentrifugation (AUC). Even covalently attached thiols desorb from the quantum dot surface in DMF, and the particles decay (reduce particle size) over time and at high temperatures [68]. Similar research involving place exchange reactions has also revealed a link. TOP/TOPO-capped CdS and CdSe quantum dots [77] as well as amine-stabilized quantum dots [78]. As a result, researchers working on quantum dots in liquid crystals should look into the effect of low quantum dot concentrations in mixtures over time, as well as changes in particle size following thermal treatments, such as during sample preparation or in liquid crystal hosts with phase transition temperatures well above ambient. Silica coatings or surface silanization could be employed to improve the stability of quantum dots in this case [79–81].

2.3 *Magnetic Nanoparticles*

Magnetic nanoparticles have attracted the attention of the liquid crystal community for a variety of reasons, including the development of magnetic nanoparticle superlattices and the construction of magnetically regulated optical liquid crystal devices. The reorientation of nematic liquid crystal molecules by an applied magnetic field, known as a Fre'edericksz transition [82, 83], has long been known, and some of the very early work on magnetic nanoparticle-doped nematic liquid crystals (so-called ferronematics) dates back to experiments performed by Brochard and de Gennes in 1970 using 7CB doped with magnetic nanoparticles with a 10 nm core [84]. The synthesis of magnetic nanoparticles is known for a variety of elements and elemental compositions, including Fe, Co, CoFe, FePt, Zn1xFexO, and others, but most magnetic nanoparticle studies, from medical applications to the liquid crystal composites discussed here, focus on commonly superparamagnetic iron oxide nanoparticles, such as magnetite (Fe_3O_4) or maghemite (γ-Fe_2O_3). Preparing magnetic iron oxide nanoparticles can be performed in a variety of ways, including co-precipitation (see below), non-aqueous and aqueous sol-gel [85], microemulsion, hydrothermal/solvothermal [86], and sonochemical techniques [87]. The co-precipitation process developed by Massart is frequently utilised due to its simplicity, with the final result typically consisting of nanoparticles ranging in size from 7 to 30 nm in diameter and including both magnetite (Fe_3O_4) and maghemite (γ-Fe_2O_3) phases. However, this approach can only be used to obtain iron oxide nanoparticles that are dispersible in water [88].

3. Quasi-Zero-Dimensional Ferroelectric Nanoparticles

Ferroelectric nanoparticles are the last type of quasi-spherical (zero-dimensional) nanoparticles to be covered in depth here. Because of the possible connection of their ferroelectric, piezoelectric, and dielectric properties, ferroelectric nanoparticles are of great interest in liquid crystal research, particularly in thermotropic nematic liquid crystal phases. $BaTiO_3$ and $Sn_2P_2S_6$ are the most studied ferroelectric nanoparticles in liquid crystals [89, 90]. However, because of the potential applications of such particles in nonvolatile memories, capacitors, and actuators, there has been a surge in the development of a variety of ferroelectric nanoparticles with compositions ranging from $PbTiO_3$ [91] and $Pb(Zr_{1-x}Ti_x)O_3$ [92], to $SrBi_2Ta_2O_9$ [93], to $Ba_{0.6}Sr_{0.4}TiO_3$ [94], and others. Apparently, some synthesis methods give real single ferroelectric domain nanoparticles, whereas others only create ferroelectric multi-domains [95]. Several high-temperature protocols (often employing autoclaves) [96], biosyntheses including template syntheses [97, 98], as well as sol-gel [99, 100], molten hydrates salt [101], and grinding processes [102] are known to make ferroelectric nanoparticles. Similar to metal nanoparticles or semiconductor quantum dots, evidence for complete coverage with surfactant molecules would be required to avoid long-term aggregation of the nanoparticles in liquid crystal phases, and should be performed using techniques described earlier in this review, such as elemental analysis in conjunction with TEM or SEM imaging.

4. One-Dimensional Nanorods

Similar to zero-dimensional metal nanoparticles, gold nanorods are the focus of practically all one-dimensional metal nanostructure research. The high interest in anisometric gold nanoclusters stems from their unique optical and electronic properties, which can be easily tuned by small changes in size, structure (for example, the position, width, and intensity of the absorption band due to the longitudinal surface plasmon resonance, which is strongly influenced by the shell as well as the aspect ratio of the nanorods), shape (e.g., needle, round capped cylinder, or dog bone), and inter-particle distance [103]. Several methods, including template-based methods [104, 105], electrochemical methods [106, 107], and seed-mediated wet chemistry methods, can manufacture gold nanorods with reasonable yields. Seed-mediated growth methods in the presence or absence of Ag^+ ions are typically used due to their simplicity and ease of nanorod size and shape control [108, 109]. This approach usually consists of three separate steps: (a) reducing $HAuCl_4$ with $NaBH_4$ to produce surfactant-capped gold nanoparticle seeds (usually ranging in diameter from 1.5 to 5 nm), (b) the production of a growth solution by reducing $HAuCl_4$ with a milder reducing agent such as ascorbic acid (both procedures in the presence of a stabilizing surfactant such as cetyltrimethylammonium bromide (CTAB)), (c) the growing solution is supplemented with the gold nanoparticle seeds generated in the first stage. The CTAB bilayer coats the gold nanorods, preventing aggregation and facilitating the creation of rather stable nanorod suspensions in aqueous solution. The CTAB bilayer coats the gold nanorods, preventing aggregation and facilitating the creation of rather stable nanorod suspensions in aqueous solution. Smith et al. have demonstrated that even minute levels of iodide present as an impurity in CTAB samples obtained from various chemical vendors had a significant impact on nanorod development. Nanorods did not form in all cases; if the iodide impurity concentration was larger than 50 ppm, only quasi-spherical nanoparticles were seen [110, 111]. When silver ions are included in the growth medium, the production of gold nanorods greatly improved. The presence of Ag^+ causes the creation of 99 percent gold nanorods, allowing for precise control of the aspect ratio between 1.5 and 4.5 and a maximum nanorod length of roughly 100 nm. It has been noticed that the tiniest seeds produce larger yields of nanorods, and that a one-step colloidal synthesis process produces the best yield and size homogeneity [112, 113]. In the presence of Ag^+ ions, a co-surfactant is required to generate greater nanorod aspect ratios and lengths; benzyldimethylhexadecylammonium chloride for aspect ratios up to 10 [114] or Pluronic F-127 for aspect ratios up to 20 [115]. In the Ag under-potential deposition mechanism [116], electric field interactions cause $AuCl_2^-$ ions on the CTAB micelles to diffuse to the CTAB-capped seed spheres, breaking the spheres symmetry into multiple facets. CTAB binding and Ag^+ deposition are faster on the 110 end facet than on the 100 end facet, resulting in preferential particle growth down the (110) axis until all silver ions have been deposited on the 100 end facet. The growth mechanism of nanorods in presence of Ag^+ is presented in Scheme 2.3 [116]. Each of the proposed mechanisms for gold nanorod growth, whether in the presence or absence of Ag^+ ions, has advantages and disadvantages, but it is widely accepted that gold nanorods synthesised using the Ag^+-assisted methodology are crystallographically different

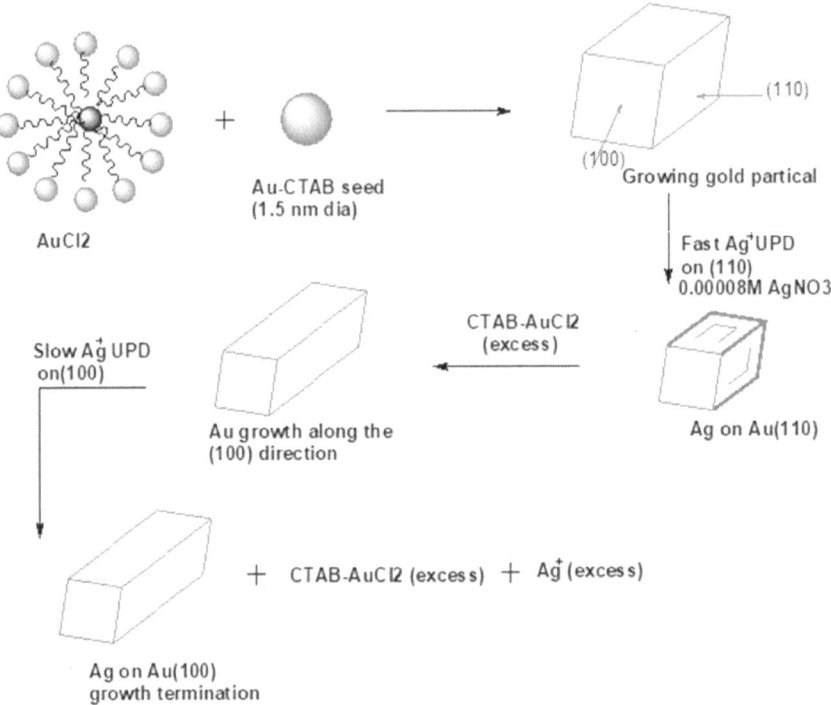

Scheme 2.3. Growth mechanism of nanorods in presence of Ag^+ ions [116].

from those synthesised without Ag^+ ions [117, 118]. According to research works on previous high-temperature syntheses, the length of the nanorods reduces as the temperature rises, but the width remains constant [113]. The aspect ratio of the resultant nanorods increases dramatically when the reaction temperature is lowered, owing to a reduction in nanorod diameter, which is explained by the confinement of diameter growth at lower temperatures [119, 120].

The formation of a considerable fraction of nanospheres and other shapes during the growth process, which can be difficult to separate from the required nanorods, is a key downside of seed-mediated growth approaches. The most typical purification method involves numerous cycles of centrifugation followed by the removal of the supernatant. Surfactant-assisted organisation of concentrated dispersions of nanorods generating ordered liquid crystalline phases, which can subsequently be precipitated out of solution, appears to be an efficient purification approach for yielding single-size nanorods from a combination of rods, spheres, and plates [121]. Even if the assumption that the heaviest nanoparticles sediment first during centrifugation is too simplistic, Sharma et al. recently demonstrated that the sedimentation behavior of gold nanorods is more strongly influenced by their diameter than by their total weight or aspect ratio, demonstrating that the key factor for sedimentation is the ratio of the nanorods' Svedberg coefficients [119].

No direct synthesis method of functionalized nanorods has been demonstrated, unlike zero-dimensional gold nanoparticles. Coating exchanges become extremely crucial for improving GNR dispersion in organic solvents, for example. The CTAB bilayer capping the nanorods is typically difficult to overcome, and further optimization of coating exchange processes looks to be a difficult undertaking that still requires major improvements [122]. The vast bulk of researches have focused on hydrophilic gold nanorods, that have been functionalized by CTAB molecules electrostatic attraction, anionic polyelectrolytes, or covalent binding of hydrophilic thiols. There are few reports on the preparation of gold nanorods with hydrophobic thiol coatings [123]. Surface modification with thiols frequently results in preferred responses on the nanorod's facet of the short axis, but not on the entire surface [124]. Similarly, thiol-for-CTAB exchanges often necessitate multiple rounds of ligand exchange to achieve near-complete hydrophobization [125, 126].

Semiconductor nanoparticles with a rod-like shape are another interesting class of one-dimensional nanomaterials that has attracted a lot of attention in the last decade because of its potential applications in electronics, optoelectronics, and sensing devices [127–129]. As compared to similar zero- or two-dimensional nanomaterials, semiconductor nanowires and nanorods have unique optical and electrical features due to their form anisotropy, such as advantageous directional capabilities in carrying electronic carriers [130, 131]. To modify these capabilities, semiconductor nanorods have been manufactured utilising a variety of methods [132–134], with core materials ranging from Si, ZnO, and ZnS to CdS and CdSe to TiO_2 and SnO_2. The vapor-liquid-solid (VLS) technique [135] is a typical method for producing nanorods from vapour, and it necessitates high operating temperatures (about 500C) [136] as well as a metal catalyst such as Au, Cu, or Ni [137]. Colloidal synthetic methods have also been employed to make semiconductor nanorods with excellent success [138]. The separation of nucleation and growth processes is a major feature of these approaches, which is often done via hot-injection [139] or heat-up protocols [140], which induce homogenous nucleation and promote uniform nanorod growth. For example, Li and coworkers devised a solvothermal synthesis [141] that uses stable inorganic precursors such nitrates and chlorides instead of the hazardous and unstable organometallic precursors used in traditional solvothermal synthetic methods [134, 142]. They were able to make CdS, ZnS, and ZnS/Mn semiconductor nanorods with adjustable aspect ratios by reacting stable inorganic salts with thioacetamide and capping them with octadecylamine or oleylamine. Template-based syntheses have also been shown to be simple and adaptable methods for generating nanorods with precise shape [134, 143, 144].

5. One-Dimensional Carbon Nanotubes

Carbon nanotubes (CNTs) have become one of the most fascinating nanomaterials in recent years [145]. CNTs are a key component for a wide range of novel applications in new materials and devices, ranging from polymer nanocomposites with enhanced mechanical strength to nanoscale electronic and optoelectronic devices [146–148]. Their unique structural, mechanical, electrical, and thermal properties [149, 150], as well as an extremely high aspect ratio, make them a key component for a wide range of novel applications in new materials and devices. There are two varieties of carbon nanotubes with high structural perfection: (a) single-walled carbon nanotubes (SWCNTs) with a typical diameter of one nanometer and (b) multi-walled carbon nanotubes (MWCNTs) with an overall outer diameter ranging from 1.4 to 100 nanometers. A SWCNT is made of a single sp^2-hybridized graphene sheet coiled into a seamless tube and sealed at both ends with a hemispherical fullerene cap. SWCNT is an uncommon one-dimensional material having a very low mass density [151] and an extraordinarily high specific surface area [152] due to its hollow structure, tiny diameter, and high aspect ratio. SWCNTs and MWCNTs are frequently produced on catalytic particles using carbon-arc discharge, laser ablation of carbon, or chemical vapour deposition [153]. CNTs, like liquid crystals, are anisotropic, which implies that their characteristics vary depending on their orientation. To acquire considerable control of the CNTs properties on a macroscopic scale and then manipulate the achieved alignment via an external field, it is necessary to regulate the orientation and specified direction of a microscopic single tube. Several studies have shown that aligned CNTs perform better in a variety of applications [154–157]. Despite the wide range of potential applications, only a few practical applications have been accomplished as yet. CNTs' usefulness is limited by critical challenges such as difficulty in manipulation and handling, mostly due to strong van der Waals pressures that cause bundling and aggregation, lack of homogeneity of physical dimensions, poor solubility, and purity control [158, 159]. Several in-growth [160, 161] and post-growth strategies for dispersion and orientation of CNTs have been developed to solve the solubility barrier and obtain aligned CNT arrays. Furthermore, there is a lot of study on how CNTs form liquid crystal phases in aqueous media [162, 163], superacids [164, 165], and DNA [166]. Among the post-growth alignment methods, suspension of CNTs in a gellan gum-based biopolymer [166] and fabrication of self-assembled systems by dispersion of MWCNTs in a nematic liquid crystalline system formed from hydrogen bonding between D-(-)-tartaric acid and undecyloxy benzoic acid [167] are two others worth mentioning. Theoretical studies of anisotropic CNTs' lyotropic or thermotropic nematic liquid crystal phase behaviour have also been published [168–170]. In terms of managing the density and form of CNTs, post-growth alignment approaches have demonstrated to be more effective than in-growth alignment methods. Nonetheless, no method can be described as adaptable in terms of achieving large-scale, high-quality alignment and dispersion. The usage of liquid crystalline materials has been a very interesting way to solving the dispersion and alignment difficulties at the same time. Liquid crystals' fluidity, and more significantly, their ability to self-organize, are helpful for CNTs as a possible solvent with improved uniform large scale alignment. It is obvious that, liquid crystal applications have also profited from the presence of CNTs.

6. Two-Dimensional Nanodiscs and Nanoclay Materials

Another very important class of nanomaterials to be studied in depth is two-dimensional, disc-shaped nanostructures. Despite the fact that a wide range of nanodiscs, nanosheets, and nanoplatelets based on metals [171], metal oxides [172, 173], graphene [174], and semiconductors [175] have been described in the literature, the vast majority of liquid crystal research have focused on nanoclay materials. Studies on the synthesis of liquid crystal phases in two-dimensional disc- or sheet-like nanomaterials, as well as the organization of nanodiscs or nanoplatelets into nematic, smectic, or columnar morphologies [176–179], are notable anomalies. The most common nanoclay material utilized in pure form or organically modified (i.e., coated with a suitable surfactant or polymer) is montmorillonite (MMT—a monoclinic smectite; having a chemical formula $(Na,Ca)_{0.3}(Al, Mg)_2Si_4O_{10}(OH)_2.nH_2O)$, beidellite $(Na_{0.5}Al_2(Si_{3.5}Al_{0.5})O_{10}(OH)_2)$.

7. Liquid Crystal Phases with Nano-particle Doping

This section will discuss on dealing with nano-particle doped nematic and smectic liquid crystal phases, considering both the achiral and chiral molecules. It will also consider the advancement of the study using other liquid crystal phases, e.g., blue phases.

7.1 Zero-dimensional Nano-particle Additives

7.1.1 Nematic Phase

The "go-to" phase for studying and understanding the major driving forces that drive interactions between suspended, quasi-spherical nanoparticles and liquid crystal molecules or combinations has unquestionably been the nematic phase. We attribute this to three key factors: (a) the availability of nematic liquid crystals, including room temperature and wide temperature range nematics, (b) the use of nematic liquid crystals in leading, large panel liquid crystal display modes such as in-plane [180, 181] and fringe-field switching [182], as well as multi-domain and patterned vertical alignment [183, 184], and (c) the use of nematic liquid crystals in leading, large panel liquid crystal display modes such as in-plane. Before entering into examples of nanoparticle-doped nematic liquid crystals, it's crucial to understand two key factors that influence the formation of quasi-spherical nanoparticles in nematic liquid crystals to some extent. The first concept is the influence of the critical nanoparticle radius on surface anchoring of nematic liquid crystal molecules on nanoparticle surfaces, as well as the anchoring energy. Both are inextricably linked to one another, and both will play a key role in a variety of nanoparticle-doped nematic systems. Planar and vertical (homeotropic) surface anchoring are the two main types of surface anchoring. Both anchoring and particle radius have a significant impact on phase-separated particle aggregation as well as local liquid crystal ordering distortion (i.e., the induction of defects around the nanoparticle). When particles agglomerate in places with the biggest liquid crystal distortions, Lavrentovich et al. formulated it qualitatively [185].

Toshima et al. created a series of monometallic Ag and Pd, as well as bimetallic Ag–Pd nanoparticles coated with 5CB with an average diameter of 1.4–3.6 nm, which reduced the phase transition temperature of the 5CB host by a maximum of 1.2C for the monometallic Ag particles (the smallest in the series being 1.4 nm in diameter), induced a frequency modulation response, and a nonlinear rising in threshold voltage. The authors propose that the observed electro-optic effects are the result of an electronic charge transfer (Coulomb attraction and repulsion) from the bimetallic nanoparticle surface via the cyano group of the 5CB capping molecules, especially if these interactions extend into the bulk via homeotropically anchored 5CB host molecules [186]. Recently, Zubarev and Link et al. described the production of very stable metal nanoparticle dispersions capped with pro-mesogenic ligands. 4-sulfanylphenyl-4-[4-(octyloxy)phenyl]benzoate-capped, mixed monolayer gold nanoparticles with a fairly narrow size distribution (around 6 nm in diameter) were prepared in a two-step process that began with a partial ligand exchange of decanethiol with 4-mercaptophenol and ended with an esterification of the phenolic hydroxy groups with 40-octyloxybiphenyl-4-carboxylic acid [187]. Yoshida et al. recently reported an alternate approach for producing stable gold nanoparticle suspensions (1–2 nm in diameter) in nematic liquid crystals [188]. They employed a simple sputter deposition method to create thin liquid crystal films with well-dispersed gold nanoparticles in both 5CB and E47 (both available from Merck), with nanoparticle sizes varying based on the nematic liquid crystal used. In nematic liquid crystals, $BaTiO_3$ particles are another popular and well-studied kind of nanoparticle. Cook et al. found that doping nematic TL205 with single domain ferroelectric $BaTiO_3$ nanoparticles (9 nm in diameter) lowered or increased the threshold voltage by 0.8 V depending on the polarity of the applied voltage [95]. Blach and coworkers also found a lower Fre'edericksz transition threshold voltage (V_{th}) for 5CB doped with large $BaTiO_3$ particles (150 nm in diameter) [189], which is surprising given an earlier report by West and Reznikov et al., who found no such reduction in V_{th} using smaller, chemically similar nanoparticles [190, 191]. In addition to lowering V_{th}, ferroelectric nanoparticles such as $BaTiO_3$ or $Sn_2P_2S_6$ [102, 192–195] have been shown to increase the nematicto-isotropic phase transition temperature ($T_{N/Iso}$) and the order parameter of the nematic host [89, 192, 196–198], which are thought to be caused by a coupling of the particles' electric dipole moment with the orientational order of the surrounding nematic molecules. Furthermore, Glaser et al. published a paper on the behaviour of spherical nanoparticles earlier this year, concentrating first on the isotropic nematogen matrix [199] and then on the nematic phase itself [200]. The nematogens demonstrate a tendency for frustrated planar anchoring in the isotropic phase, which results in a long-range diminution of the orientational ordering relative to the pure isotropic phase. The size of suspended nanoparticles was shown to affect the local orientational order and weakly long-range repulsive interactions between them [199]. Glaser et al. found that spherical nanoparticles are more "soluble" in the isotropic phase than in the nematic phase, confirming previous experiments. Finally, for the case of homeotropic anchoring, nematic matrix induced intermediate-range repulsion interactions between nanoparticles and nematic molecules in the isotropic phase were described, but particle–particle interactions not only suppressed repulsion but also caused nanoparticle aggregation [200]. Kitzerow et al. showed that temperature-induced phase transitions (Iso-N) and electric field-induced reorientation of a nematic liquid crystal (5CB in this case) can be used to tune photonic modes of a microdisc resonator [201], in which embedded InAs quantum dots serve as emitters feeding the optical modes of the GaAs-based photonic cavity. Tong and Zhao proposed modulating the photoluminescence intensity of CdSe/ZnS core-shell quantum dots using the electric field response of a chiral nematic liquid crystal comprising either a self-assembled physical network or a covalently cross-linked polymer [202]. The authors used low concentrations of quantum dots (0.02 or 0.2 wt percent) to avoid aggregation in these networks, which could explain why no changes in electro-optic characteristics were detected. For example, Zhang et al. observed that 5CB doped with CdS quantum dots had lower threshold voltages (by up to 25% in twisted nematic cells) [203]. When combined with data on insulating nanoparticles like ZnO, TiO_2, and Si_3N_4 in nematic liquid crystals [204], it's clear that most metallic, semiconducting, ferroelectric, and insulating nanoparticles can be used to tune and improve the electro-optic properties of nematic liquid crystals without the need for additional chemical synthesis. The most noticeable and promising trait related to applications in display devices is likely a drop in the operational or threshold voltage, which is not observed for all systems [205], but the explanations for this reduction in Vth are as diverse as the colloids utilized in these investigations. Segregation of nanoparticles at liquid crystal/

substrate interfaces can also help promote or change the alignment of thin nematic liquid crystal films. Several types of nanoparticles have been demonstrated to generate homeotropic alignment as well as defects and remarkable defect patterns, depending on their concentration, size, and nature. Nanoparticle analogues such as polyhedral oligomeric silsesquioxanes (POSS) have been found to induce homeotropic alignment in ITO/glass cells without alignment layers and in so-called hybrid aligned nematic (HAN) cells, according to Hwang et al. and Liao et al. [206, 207]. Masutani et al. demonstrated that nanoparticles can be inserted into polymer dispersed liquid crystal (PDLC) matrices to reduce gain reflector viewing angle dependency and metallic glare [208]. Huang et al. have shown that negatively charged silica nanoparticles (Aerosil®R812, Degussa-Huls, 7 nm diameter) migrate to the planar side of a HAN cell after being driven with a d.c. pulse voltage and induce homeotropic alignment (a memory mode). In concept, the authors of the latter paper [209] created a dual mode device with a memory mode after applying a d.c. pulse and a dynamic mode with faster response times when the same cell is driven with an a.c. pulse.

7.1.2 *Chiral Nematic Phase*

For nanoparticle-doped chiral nematic liquid crystals, some intriguing phenomena described for non-chiral nematic systems were also reported. We discussed the work of Kobayashi et al., who used metal nanoparticles as dopants to demonstrate a frequency modulation twisted nematic (FM-TN) mode and quick switching characteristics. Guo et al. presented an intriguing concept based on magnetite (Fe_3O_4) nanoparticles that were either bare (about 50 nm in diameter) or coated with oleic acid (smaller in diameter). Unfortunately, the specific size and its distribution, as well as the exact nature of the iron and iron oxide surface states, were not revealed. Nonetheless, the authors demonstrated that oleic acid-protected magnetite nanoparticles formed more homogeneous dispersions than bare magnetite particles, and that composites of magnetite nanoparticles capped with oleic acid and further modified with a chiral dopant can be developed into flexible and magnetically addressable/erasable colour optical devices in the future [210]. Qi et al. took a different strategy, combining nanoparticle characteristics with chiral nematic liquid crystal phases [211]. Their plan was to coat gold nanoparticles with chiral compounds that are known to generate chiral nematic phases strongly. We created a series of alkylthiol-capped gold nanoparticles, either pure monolayer or mixed monolayer, with all or almost all of the alkylthiols end-functionalized with (S)-naproxen to implement the concept.

7.1.3 Smectic Phases

Compared to nematic liquid crystals, smectic liquid crystals doped with quasi-spherical nanoparticles have become increasingly difficult to come by in recent years. This is surprising, especially in light of recent research by Smalyukh et al., who discovered that nanoscale dispersion (based on N-vinyl-2-pyrrolidone-capped gold nanoparticles with 14 nm diameter) in a thermotropic smectic liquid crystal (8CB) is potentially much more stable than nanoparticle dispersions in nematics [212]. Over the last two decades, researchers have been studying how hydrophilic and hydrophobic Aerosil™ particles form gels and networks in nematic and smectic liquid crystal phases. For example, Nounesis et al. investigated both hydrophilic Aerosil™ 300 and hydrophobic Aerosil™ R812 (both with a diameter of approximately 7 nm, Degussa Corp.) in three different SmA-SmC(*) materials (two chiral materials with a SmA-SmC* and one achiral SmA-SmC compound) in three different SmA-SmC(*) materials (two chiral materials with an SmA-SmC* and one achiral SmA-SmC compound). The research group confirmed that the layer thickness in the SmA phase decreased monotonically as the concentration of Aerosil™ particles increased, and that this effect was more pronounced for the achiral liquid crystal host, using high resolution calorimetric and small angle X-ray diffraction experiments. The layer shrinkage showed a strong crossover in both composites based on chiral hosts, where the Aerosil™ network revealed its concentration-dependent soft-stiff transition. The layer contraction, according to the authors, is linked to an aerosil-induced tilt of the LC molecules with respect to the layer planes [213]. Mart'nez-Miranda et al. investigated the effects of different coatings on the surface of magnetic Fe_xCo_y nanoparticles (2 and 11 nm in diameter) in the SmA phase of 8CB, and found that the coating nature had a significant impact on the reorientation of the liquid crystal molecules in response to an applied magnetic field [214].

7.1.4 Blue Phases

Blue phases have mostly remained a source of scientific fascination in the past since, for many years, the relatively narrow temperature range (about 0.5 to 2°C) severely hindered future scientific research as well as practical use of these fascinating phases. New materials [215–221], their reaction to electric fields [222–229], the morphology of blue phase crystals generated, for example, by an applied electric field [230, 231], and lasing in blue phases [232] were the main topics of early research on blue phases. Blue phases are now undergoing a renaissance [233, 234] with the finding of wide temperature range blue phases in liquid crystal bimesogens (or terminal twins) by Coles et al. [235] and polymer-stabilized blue phases by Kikuchi et al. [236]. Blue phases with a broad temperature range (including room temperature) are now sought after and being produced for use in display applications [237–239], lasing [240, 241], and tunable photonic band gap materials [242, 243]. It was just a matter of time before blue phase researchers started doping them with quasi-spherical nanoparticles. Although this field is still in its infancy, recent publications have shown some encouraging results. Yoshida et al., for example, reported a drop in the clearing point of roughly 13°C and an expansion of the temperature range of cholesteric blue phases from 0.5 to 5°C by doping blue phases with gold nanoparticles (average diameter of 3.7 nm) [244].

7.1.5 Columnar Phase

Surprisingly, columnar liquid crystal phases have received little consideration as nanoparticle host materials. Columnar liquid crystal phases have excellent charge-carrier mobilities and generation efficiencies, making them ideal candidates for organic photovoltaics [245, 246], and these properties might be changed, if not improved, by adding modest amounts of metal, semiconductor, or other nanoparticles. The increased viscosity of particular columnar phases (relative to nematics), the reduced dispersibility of columnar phases as a result, and the general shape mismatch between quasi-spherical nanoparticles and disc-like molecules stacking to form columns could all be factors in the development of such composites. Kumar et al. were able to create stable suspensions of hexanethiol-capped gold nanoparticles (1.6 nm core diameter) and gold nanoparticles capped with a 10CB thiol in both a lyotropic hexagonal columnar phase (H1 phase formed by a 42:58 w/w Triton X-100/water system) and an inverse hexagonal columnar phase (H_2 phase formed by AOT). In a hexagonal columnar triphenylene host, the same research group reported on dispersions of gold nanoparticles capped with triphenylene-based discotic liquid crystals, and discovered that doping the columnar phase with these nanoparticles increased the electrical conductivity by several orders of magnitude [247, 248]. In the same hexagonal columnar liquid crystal doped with methylbenzene thiol-protected gold nanoparticles, Holt and co-workers demonstrated a similarly substantial increase (106) in electrical conductivity (diameter of the core: 2.7 nm). This conductivity boost could be due to nanoparticle aggregation and the development of smaller chain-like aggregates, according to the authors [249].

7.2 One-Dimensional Nanoparticle Additives

7.2.1 Metal Nanorods

The use of nanorods, nanotubes, and nanowires as nonlinear optical (NLO) materials, photo-sensors [250], photo-induced memory devices, and a variety of other technologically significant applications has grown in popularity over the last decade. The necessity for ordered and assembled nanorod designs has prompted the creation of novel approaches for organising these anisotropic building elements. (i) embedding metal nanorods in a polymer film and then stretching the composite film (with the nanorods tending to align in a preferred orientation dictated by the elongation of the polymer molecules) [251, 252], and (ii) nanorod surface chemistry-based alignment and assembly methods. The use of nanorods, nanotubes, and nanowires as nonlinear optical (NLO) materials, photo-sensors [250], photo-induced memory devices, and a variety of other technologically significant applications has grown in popularity over the last decade. Long-range nanorod alignment could be achieved by doping nanorods into liquid crystals, which could be an alternate method. Furthermore, tunable metamaterials could be created by aligning liquid crystal molecules on substrates and reorienting them in response to an applied electric or magnetic field. However, in practise, this has proven to be a challenging undertaking, owing to issues with creating thermotropic LC mixes with well dispersed GNRs [253]. Smalyukh et al. [254] have published a work on nanorod alignment utilising potentially more suited lyotropic liquid crystals. The authors exhibited spontaneous long-range orientational ordering of CTAB-capped GNRs dispersed in the CTAB/benzyl alcohol/water ternary system's lyotropic nematic and hexagonal columnar liquid crystalline phases. Bulk samples with an area of roughly 1 in^2 and thicknesses ranging from a few micrometres to millimetres were used to achieve the alignment. Unidirectional alignment of well scattered GNRs can be shown in freeze-fracture TEM images corresponding to different fracture planes containing the director. The nanorods are separated by distances similar to their length and followed the direction of the anisotropic fluids' director. There was no correlation between the nanorods' centres of mass, which is similar to common nematics. Based on these TEM images, the two-dimensional orientational order parameter was determined to be 0.86–0.91, whereas the three-dimensional ordering was derived from polarised extinction spectra acquired at various angles between the linear polarizer and the director. A three-dimensional order parameter of a shear-aligned columnar hexagonal sample was calculated to be around 0.46 by the authors. This low number is owing to the director's imprecise alignment at the millimetre scale, which revealed micron-sized domains with slightly misaligned director orientations. It was demonstrated using dark-field microscopy that the local order is higher within these microdomains and that the order parameter values are compatible with those determined from TEM pictures. Furthermore, when a magnetic field is applied to a nematic sample that has been annealed from its isotropic phase, both the rod-like micelles and suspended nanorods orient along the external magnetic field and maintain this uniform alignment when the field is removed. The GNRs were imposed to follow the director of the liquid crystal, and the orientation of GNRs in the nematic phase can be altered by magnetic fields, according to polarised surface plasmon resonance (SPR) spectra of these magnetically aligned samples. Prasad et al. investigated the electro-optic characteristics of GNR/8CB suspensions [255], interested in the electrical impact of GNR on nematic liquid crystals. The inclusion of rod-like colloidal gold particles into the nematic fluid (i) did not destabilise the mesophase, (ii) raised the dielectric anisotropy by roughly 14%, and (iii) greatly increased the elasticity anisotropy, according to the researchers. Furthermore, GNR doping increased both $\sigma_{\parallel}$ and $\sigma_{\perp}$, resulting in a three-fold increase in electrical conductivity. However, unlike quasi-spherical gold nanoparticles, the inserted GNRs did not appreciably alter the conductivity anisotropy, most likely due to their anisometric form. Electro-optic effects induced by doping liquid crystals with one-dimensional metal nanoparticles were investigated not only in standard electro-optic test cells, but also in costume-made cells consisting of a thin layer of liquid crystal deposited onto a thin film of alumina with embedded GNRs [256], or using rubbed polyimide alignment layers modified with solution-cast GNR [257]. Electrical modulation of the surface plasmon resonance frequencies of the GNR integrated into these liquid crystal cells was possible in both circumstances. Finally, controlled spatial alterations in the alignment of nematic liquid crystals can be used to manipulate and organise magnetic nanowires and other submicrometer-scale anisometric particles. For example, Leheny and colleagues measured the elastic strains placed on anisotropic Ni nanowires hanging in a nematic liquid crystal and demonstrated that introducing a magnetic field causes the nanowire to reorient and deform the director in the surrounding area [258].

7.2.2 *Semiconductor Nanorods*

One-dimensional semiconductor nanorods are interesting dopants for liquid crystals due to their unique electrical, optical, and mechanical capabilities. As a result, composites of these two dissimilar materials have a lot of potential for use in electrical and optical devices. Chen et al. described a broad method for controlling the polarisation of emission from semiconductor nanorods using an external bias. To achieve the maximum polarisation ratio of the suspended nanorods, they employed a composite of a nematic liquid crystal mixture (E7, Merck) and nanorods (CdS) poured into an ITO-coated cell with an optimal concentration of one CdS nanorod per 10^{10} LC molecules [259, 260]. In this technique, a nematic liquid crystal serves as a solvent and medium, with the direction of alignment controlled by an applied electric field. As a result of the anchoring force between the liquid crystal molecules and the nanorods, the orientation of the CdS nanorods can be fine-tuned by an external bias. This one-of-a-kind trait of fine-tunable emission anisotropy opens up new avenues for smart optoelectronic device applications.

Liquid crystals' electro-optic characteristics were also modified via interactions between nanorods and liquid crystals. In nematic liquid crystals doped with nanorods, several groups observed decreased threshold voltages [261–264]. The long range ordering of a blend of ultra-narrow ZnS nanorods (1.2 nm in diameter and 4.0 nm in length) in a nematic liquid crystal mixture (ZLI-4792) was enhanced due to a favourable energetic configuration as a result of the liquid crystal's refractive index compatibility and the octadecylamine surfactant coating the nanorods, according to Acharya et al. [261, 262]. Furthermore, the effective dielectric anisotropy arising from the strong intrinsic dipole moment (37 D) of these ultrasmall nanorods improved the electro-optic characteristics of this blend. In compared to pristine liquid crystal, the blend's threshold voltage was reduced by up to 10%, and the optical response time was improved to 4 ms from 33 ms for the plain liquid crystal mixture. On a related topic, Ouskova and colleagues reported that a composite of magnetic β-Fe$_2$O$_3$ nanorods in 5CB had lower threshold voltages than pure 5CB and that the sensitivity of the nematic liquid crystal to external magnetic fields was increased in the presence of such magnetic nanorods [263]. Finally, several groups interested in the macroscopic organisation and orientation of nanorods reported on the formation of a lyotropic liquid crystal phase induced by the self assembly of polymer-coated semiconductor nanorods [264–268], which could be used to improve device performance in solar cells, for example.

7.2.3 Carbon Nanotubes

The self-organization of thermo tropic and lyotropic liquid crystals makes them exceptional mediums for the dispersion and organisation (alignment) of carbon nanotubes. Scalia's recent excellent review [158], Popa-Nita and Kralj's theoretical work on anchoring at the liquid crystal/CNT interface [269], and a number of earlier experimental reports on liquid crystal/CNT composites demonstrating that liquid crystal orientational order can be transferred to dispersed CNTs, which is commonly illustrated using polarised Raman spectroscopy [270–273], have all focused on this topic. CNTs integrate into liquid crystal matrices and are capable of producing noticeable changes on the electro-optic characteristics of liquid crystals, even at extremely low concentrations, as shown by this collaborative work, which could be very appealing for display applications. As a result, the quality of CNT dispersion in liquid crystals is an incredibly significant issue. Lagerwall et al. extensively investigated numerous CNT combinations utilising a variety of pure and multicomponent liquid crystals to investigate this, i.e., limiting aggregation and boosting the long-term stability of dispersions [274]. Lagerwall and Tschierske et al. previously demonstrated that adding polyphilic liquid crystalline molecules that act as dispersion promoters to liquid crystal/CNT dispersions is a viable and promising method for stabilising CNT [275]. In this study, three different sub-units were synthesised: a rod-like mesogenic cyanobiphenyl core that is compatible with the liquid crystal host, flexible hydrocarbon or ethylene oxide spacers, and terphenyl, anthracene, or pyrene anchor groups that interface with the CNTs via p–p stacking. Kimura and co-workers [276] described another interesting method for arranging CNTs using liquid crystals. The researchers created p-conjugated oligo(phenylene vinylene) (OPV) liquid crystals with terminal cyanobiphenyl moieties and demonstrated the formation of stable supramolecular complexes with SWCNTs via strong p–p interactions between the tubes' sidewalls and the phenylene vinylene segments. These OPV units can also wrap around SWCNTs, giving them high organic solvent solubility. Several additional groups [277, 278] presented noteworthy experimental and theoretical results on CNT alignment and electro-optic response in thermotropic nematic liquid crystals. Lyotropic liquid crystals, on the other hand, should be considered as a possible platform for CNT dispersion and alignment. Many flexible CNT dispersants [279] may generate lyotropic liquid crystalline phases, and successful CNT dispersion and alignment in aqueous media can open new possibilities for biological applications [280]. SDS, the most popular surfactant for dispersing and aligning CNTs, can be used to disperse commercial HiPCO SWCNTs (from a high-pressure CO conversion procedure) without any optically evident bundling in its nematic liquid crystal phase with disc- or rod-shaped micelles [281, 282]. Surfactants (and the lyotropic liquid crystal phases created by these molecules) have the intrinsic ability to keep CNTs properly dispersed, thus counteracting CNTs' strong desire to congregate into disorderly bundles, as well as macroscopically uniform alignment. Integration of SWCNTs in a hexagonal lyotropic liquid crystal phase has been successfully demonstrated in addition to nematic lyotropic phases [283–285]. Using small angle X-ray scattering experiments, Regev et al. found that SWCNTs can be integrated into such liquid crystal phases, as evidenced by an increase in the lattice parameter at CNT concentrations up to 0.15 wt%, which the authors attribute to a swelling induced by the incorporation of SWCNTs within the hexagonal phase's cylinders. A reduction in the lattice parameter above 0.15 wt percent was thought to indicate slow phase separation [283]. Lagerwall et al. went on to show that combining cationic and anionic surfactants to form a hexagonal columnar lyotropic liquid crystal phase not only produces a well-dispersed, macroscopically aligned suspension of unsupported particles, but also generates a well-dispersed, macroscopically aligned suspension of supported particles. CNTs with a 20-fold increase in nanotube concentration, as well as the ability to transport liquid crystal-dispersed CNTs to solid surfaces and access to aligned thin filaments, all of which could prove to be excellent methods for handling and processing. CNT processing [284, 285]. This method could pave the way for new approaches to fabricate devices that require CNTs to be put on substrates and aligned in specific directions. Similarly, various investigations have been conducted with the goal of improving the electro-optic properties of liquid crystals by introducing carbon nanotubes. Several recent papers [286–290] have described the synthesis, characterisation, and study of electro-optic properties of liquid crystal/CNT dispersions using widely used nematic liquid crystals, which have been reviewed by Lee et al. [291]. CNTs, for example, produced a dielectric permittivity in 5CB that was frequency and concentration dependent [292]. CNTs have a considerable impact on the isotropic to nematic phase transition in CNT-doped liquid crystals, according to Basu et al. [293].

Electro-optic and dielectric characteristics of ferroelectric liquid crystals doped with chiral CNTs were explored by Haase and colleagues [294, 295]. Even a minor concentration of CNTs had a significant impact on the performance of the doped liquid crystal mixture. The experimental results were explained by two effects: (i) the p-electron system of the CNT screens spontaneous polarisation of the ferroelectric liquid crystal, and (ii) the CNT p-electrons trap ionic impurities, resulting in a significant change in the internal electric field within liquid crystal test cells. Lagerwall and Dabrowski et al. [296] showed another example of SWCNT dispersion in a multi-component antiferroelectric mectic-C* liquid crystal mixture. In contrast to the undoped sample, which exhibited a SmA and two unique intermediate phases, a SmC*β and a SmC*γ phase, SWCNTs produced the emergence of a single-layer SmC* phase between the SmA phase and the crystalline state in this work. Finally, MWCNT dispersions in chiral nematic liquid crystals were investigated. The helical twisting features of the chiral nematic phase did not alter as a result of our tests. The MWCNTs, on the other hand, were assumed to disrupt translational order in the SmA phase (decrease of the SmA-N* phase transition) while following the chiral nematic phase twist of the nematic direction [297].

7.2.4 Two Dimensional Nanoclay

Organically modified clay suspensions in liquid crystal phases nearly exclusively target the nematic host 5CB. Exfoliation (diffusion in the nematic host and penetration of nematic molecules between clay sheets), aggregation, and induced electro-optic effects are often studied using calorimetric measurements, small and broad angle X-ray diffraction, as well as optical and electro-optic analyses. Pizzey et al. earlier work supplied part of the groundwork for subsequent research. They discovered that laponite clay stabilised with a quaternary ammonium surfactant generates relatively stable suspensions in 5CB that consists of freely tumbling discs that do not agglomerate. Claytone® AF, on the other hand, created disc stacks in the same liquid crystal host [298, 299]. According to van Duijneveldt and co-workers' optical microscopy and small angle X-ray scattering inspections, there is a noticeable variation in how certain clays behave in 5CB as a nematic host in terms of their dispersion or aggregation propensity [300, 301]. Huang et al., as well as Puchkovska and coworkers, focused on the electro-optic effects generated by the addition of organoclays to 5CB in a series of research. Puchkovska et al. reported electro-optic contrast and memory effects as a result of van der Waals interactions between the two components, as well as interface orientational effects [302–304], and Huang et al. concluded that the underlying cause for these observed electro-optic effects is the effective trapping of impurity ion charges that enhances the nematic molecular order [305–307]. For polymer-dispersed liquid crystal (PDLC) devices containing organically modified clays, optical transparency and memory effects (i.e., a residual transparency after the electric field is switched off) were also described [308]. Finally, Lee et al. [309, 310] were successful in creating holographic PDLC devices made of either pure or organically modified montmorillonite clay.

8. Conclusions

Over the last decade, the discipline of liquid crystal nanoscience has become extremely popular. Nanoparticle-doped liquid crystals research employs a wide range of nanoparticles in varying sizes, shapes, core materials, and coatings. The most substantial development has been made for nanoparticle-doped nematic phases, both chiral and achiral, and the reported alignment and electro-optic effects promise well for their employment in electro-optic applications. Although there has yet to be a substantial breakthrough in smectic liquid crystal phases, preliminary studies in the area of nanoparticle-doped ferroelectric liquid crystals seems promising. Based on the growing interest in blue phase mode displays, and if stable nanoparticle-doped blue phase composites can be created that show both wide temperature ranges and lower operating voltages (here Kerr effect) than nanoparticle-doped nematic liquid crystal composites (Fre'dericksz effect), blue phases will undoubtedly see a surge in research over the next few years. Many different nanoparticle core materials, shapes, and surface functionality combinations have yet to be investigated independently of the liquid crystal host phase, and more focused structure/size/shape-property relationship studies are needed to draw comprehensive conclusions about nanoparticle-induced effects. Such experiments should preferably begin with completely monodisperse nanoparticles (so-called magicnumbered or magic-sized) and adjust only one parameter at a time (e.g., the hydrocarbon chain length of the capping ligands). The long-term stability of the nanomaterials utilised, as well as the liquid crystal composites incorporating them, are other key issues that will require further investigation in the future. Similarly, given the plethora of possible differences from one batch to the next, a thorough characterization of all nanoparticles used is required, including size, size distribution, surface coverage, thermal stability, and ligand distribution on the nanoparticle surface (for mixed monolayercapped nanoparticles), to name a few.

Finally, the self-assembly of nanoparticles covered with mesogenic and pro-mesogenic capping agents is predicted to see tremendous expansion in the next years. This paper summarises a variety of various methodologies, demonstrating the creation of nematic, smectic-like, cubic, and columnar phases and/or superstructures. Again, the possibilities for creating metamaterials with nanoparticles and liquid crystal motifs are limitless, and future research will undoubtedly uncover new, more complicated phase morphologies as well as highly controllable nanoscale features as a result of liquid crystal phase development.

References

[1] Collings, P.J. and M. Hird. 1997. *Introduction to Liquid Crystals Chemistry and Physics*. Taylor & Francis. doi:10.1080/10587250008031049.

[2] Chandrasekhar, S. 1992. *Liquid Crystals*. Cambridge University Press: Cambridge,UK. doi:10.1017/CBO9780511622496.

[3] Singh, S. 2002. *Liquid Crystal Fundamentals*. World Scientific: Singapore.

[4] Jones, R.A.L. 2020. *Soft Condensed Matter*. Oxford University Press.

[5] Hamley, E.W. 2007. *Introduction to Soft Matter*. Revised Edition. Wiley. Chichester, UK.

[6] Kleman, M. and O.D. Lavrentovich. 2003. *An Introduction Soft Matter Physics*. Springer: Berlin, Germany.

[7] Hirst, L.S. 2013. *Fundamentals of Soft Matter Science*. CRC Press: Boca Raton, FL, USA.

[8] Terentjev, E.M. and D.A. Weitz. 2015. *The Oxford Handbook of Soft Condensed Matter*. Oxford University Press: Oxford, UK.

[9] Dierking, I., G. Scalia, P. Morales and D. LeClere. 2004. Aligning and reorienting carbon nanotubes with nematic. *Liquid Crystals* 16: 865–869.

[10] Dierking, I., G. Scalia and P. Morales. 2005. Liquid crystal–carbon nanotube dispersions. *J. Appl. Phys.* 97: 044309.

[11] Lagerwall, J., G. Scalia, M. Haluska, U. Dettlaf-Weglikowska, S. Roth and F. Giesselmann. 2007. Nanotube alignment using lyotropic liquid crystals. *Adv. Mater.* 19: 359–364.

[12] Kumar, S. and H.K. Bisoyi. 2007. Aligned carbon nanotubes in the supramolecular order of discotic liquid crystals. *Angew. Chem.* 46: 1501–1503.

[13] Song, W., I.A. Kinloch and A.H. Windle. 2003. Nematic liquid crystallinity of multiwall carbon nanotubes. *Science* 302: 1363.

[14] Badaire, S., C. Zakri, M. Maugey, A. Derre, J.N. Barisci, G. Wallace and O. Poulin. 2005. Liquid crystals of DNA-stabilized carbon nanotubes. *Adv. Mater.* 13: 1673–1676.

[15] Kim, J.E., T.H. Han, S.H. Lee, J.Y. Kim, C.W. Ahn, J.M. Yun and S.O. Kim. 2011. Graphene oxide liquid crystals. *Angew. Chem.* 50: 3043–3047.

[16] Xu, Z. and C. Gao. 2011. Aqueous liquid crystals of graphene oxide. *ACS Nano.* 5: 2908–2915.

[17] Shen, T.Z., S.H. Hong and J.K. Song. 2014. Electro-optical switching of graphene oxide liquid crystals with an extremely large Kerr coefficient. *Nat. Mater.* 13: 394.

[18] Al-Zangana, S., M. Iliut, M. Turner, A. Vijayaraghavan and I. Dierking. 2016. Properties of a thermotropic nematic liquid crystal doped with graphene oxide. *Adv. Mater.* 4: 1541–1548.

[19] Ren, Z., C. Chen, R. Hu, K. Mai, G. Qian and Z. Wang. 2012. Two-step self-assembly and lyotropic liquid crystal behavior of TiO_2 nanorods. *J. Nanomater.* 2012: 180989.

[20] Li, L.S., J. Walda, L. Manna and A.P. Alivisatos. 2002. Semiconductor nanorod liquid crystals. *Nano Lett.* 2: 557–560.

[21] Thorkelsson, K., P. Bai and T. Xu. 2015. Self-assembly and applications of anisotropic nanomaterials. A review. *Nano Today.* 10: 48–66.

[22] Al-Zangana, S., M. Turner and I. Dierking. 2017. A comparison between size dependent paraelectric and ferroelectric $BaTiO_3$ nanoparticle doped nematic and ferroelectric liquid crystals. *J. Appl. Phys.* 121: 085105.

[23] Mertelj, A., D. Lisjak, M. Drofenik and M. Copic. 2013 Ferromagnetism in suspensions of magnetic platelets in liquid crystal. *Nature.* 504: 237.

[24] Liu, Q.K., Y.X. Cui, D. Gardner, X. Li, S.L. He and I.I. Smalyukh. 2010. Self-alignment of plasmonic gold nanorods in reconfigurable anisotropic fluids for tunable bulk metamaterial applications. *ACS.* 10: 1347–1353.

[25] Dintinger, J., B.J. Tang, X.B. Zeng, F. Liu, T. Kienzler, G.H. Mehl, G. Ungar, C. Rockstuhl and T. Scharf. 2013. A self-organized anisotropic liquid-crystal plasmonic metamaterial. *Adv. Mater.* 25: 1999–2004.

[26] Lehmann, M. and M. Huegel. 2015. A perfect match: fullerene guests in star-shaped oligophenylenevinylene mesogens. *Angew. Chem.* 54: 4110–4114.

[27] Hu, T., Y. Gao, Z.L. Wang and Z.Y. Tang. 2009. One dimensional self-assembly of inorganic nanopartical. *Front Phys China* 4: 487.

[28] Galian, R.E. and de la M. Guardia. 2009. Liquid crystal-material design and self-assembly. *Trac-Trend Anal. Chem.* 28: 279.

[29] Reiss, P., M. Protiere and L. Li. 2009. Liquid crystal-material design and self-assembly. *Small.* 5: 154.

[30] Khanna, V.K. 2008. Liquid crystal-material design and self-assembly. *Defence Sci. J.* 58: 608.

[31] Stroyuk, A.L., V.V. Shvalagin, A.E. Raevskaya, A.I. Kryukov and S.Y. Kuchmii. 2008. Photochemical formation of semiconducting nanostructures. *Theor. Exp. Chem.* 44: 205.

[32] Nune, S.K., P. Gunda, P.K. Thallapally, Y.Y. Lin, M.L. Forrest and C.J. Berkland. 2009. Nanoparticles for biomedical imaging. *Expert Opin. Drug Del.* 6: 1175.

[33] Peng, XH., J.Y. Chen, J.A. Misewich and S.S. Wong. 2009. Carbon nanotube–nanocrystal heterostructures. *Chem. Soc. Rev.* 38: 1076.

[34] Shchukin, D.G., D. Radziuk and H. Mohwald. 2010. Feedback active coatings with sensitive containers based on nano-, micro-, and macroemulsions of direct or inversed type. *Annu Rev. Mater Res.* 40: 345.

[35] Bai, X.T., X.W. Li and L.Q. Zheng. 2010. Liquid crystal-material design and self-assembly. *Langmuir* 26: 12209.

[36] Safavi, A. and S. Zeinali. 2010. Synthesis of highly stable gold nanoparticles using conventional and geminal ionic liquids. *Colloid Surface A* 362: 121.

[37] Ballarin, B., M.C. Cassani, D. Tonelli, E. Boanini, S. Albonetti, M. Blosi and M. Gazzano. 2010. Gold nanoparticle-containing membranes from *in situ*

reduction of a gold(III)–Aminoethylimidazolium aurate salt. *J. Phys. Chem. C* 114: 9693.

[38] Obliosca, J.M., I.H.J. Arellano, M.H. Huang and S.D. Arco. 2010. Morphogenesis of anisotropic gold nanostructures stabilized by the greener ionic liquid 1-butyl-3-methylimidazolium lauryl sulfate. *Mater Lett.* 64: 1109.

[39] Brust, M., J. Fink, D. Bethell, D.J. Schiffrin and C. Kiely. 1995. Synthesis and reactions of functionalised gold nanoparticles. *J. Chem. Soc. Chem. Commun.* 1655.

[40] Brust, M., M. Walker, D. Bethell, D.J. Schiffrin and R. Whyman. 1994. Synthesis of thiol-derivatised gold nanoparticles in a two-phase liquid–liquid system. *J. Chem. Soc. Chem. Commun.* 801.

[41] Brust, M., D. Bethell, C.J. Kiely and D.J. Schiffrin. 1998. Colloids and colloid assemblies. *Langmuir* 14: 5425.

[42] Bethell, D., M. Brust, D.J. Schiffrin and C. Kiely. 1996. Nanoparticle assemblies superstructures. *J. Electroanal. Chem.* 409: 137.

[43] Goulet, P.J.G. and R.B. Lennox. 2010. Functionalized nanomaterials. *J. Am. Chem. Soc.* 132: 9582.

[44] Kassam, A., G. Bremner, B. Clark, G. Ulibarri and R.B. Lennox. 2006. Place exchange reactions of alkyl thiols on gold nanoparticles. *J. Am. Chem. Soc.* 128: 3476.

[45] Zachary, M. and V. Chechik. 2007. Liquid crystal-Material design and self-assembly. *Angew Chem. Int. Ed.* 46: 3304.

[46] Rucareanu, S., V.J. Gandubert and R.B. Lennox. 2006. Liquid crystal-material design and self-assembly. *Chem. Mater.* 18: 4674.

[47] Gandubert, V.J. and R.B. Lennox. 2005. Liquid crystal-material design and self-assembly. *Langmuir* 21: 6532.

[48] Weare, W.W., S.M. Reed, M.G. Warner and J.E. Hutchison. 2000. Improved synthesis of small (dCORE ≈ 1.5 nm) phosphine-stabilized gold nanoparticles. *J. Am. Chem. Soc.* 122: 12890.

[49] Dasog, M. and R.W.J. Scott. 2007. Liquid crystal-material design and self-assembly. *Langmuir* 23: 3381.

[50] Hou, W.B., M. Dasog and R.W.J. Scott. 2009. Probing the relative stability of thiolate- and dithiolate-protected Au monolayer-protected clusters. *Langmuir* 25: 12954.

[51] Shon, Y.S., S. Chuc and P. Voundi. 2009. Stability of tetraoctylammo-nium bromide-capped gold nanoparticles: effects of anion treatments. *Colloid Surface A* 352: 12.

[52] Joseph, Y., I. Besnard, M. Rosenberger, B. Guse, H.G. Nothofer, J.M. Wessels, U. Wild, A. Knop-Gericke, D.S. Su, R. Schlogl, A. Yasuda and T. Vossmeyer. 2003. Handbook of gas sensor materials. *J. Phys. Chem. B* 107: 7406.

[53] Bang, J., J. Park, J.H. Lee, N. Won, J. Nam, J. Lim, B.Y. Chang, B. Chon, J. Shin, J.B. Park, J.H. Choi, K. Cho, S.M. Park, T. Joo and S. Kim. 2010. Semiconductor quantum dots-organometallic and inorganic system. *Chem. Mater.* 22: 233.

[54] Zhang, W.J., G.J. Chen, J. Wang, B.C. Ye and X.H. Zhong. 2009. Liquid crystal-material design and self-assembly. *Inorg. Chem.* 48: 9723.

[55] Yang, Y.A., O. Chen, A. Angerhofer and Y.C. Cao. 2009. Radial-position-controlled doping of CdS/ZnS core/shell nanocrystals: surface effects and position-dependent properties. *Chem-Eur. J.* 15: 3186.

[56] Yang, H.W., W.L. Luan, Z. Wan, S.T. Tu, W.K. Yuan and Z.M. Wang. 2009. Liquid crystal-material design and self-assembly. *Cryst. Growth Des.* 9: 4807.

[57] Law, W.C., K.T. Yong, I. Roy, H. Ding, R. Hu, W.W. Zhao and P.N. Prasad. 2009. Smart material and nanotechnology in engineering. *Small* 5: 1302.

[58] Bhattacharyya, S., Y. Estrin, O. Moshe, D.H. Rich, L.A. Solovyov and A. Gedanken. 2009. Liquid crystal-material design and self-assembly. *ACS Nano* 3: 1864.

[59] Sharma, M., D. Gupta, D. Kaushik, A.B. Sharma and R.K. Pandey. 2008. Highly luminescent inverted ZnS/CdS core/shell quantum dots. *J. Nanosci. Nanotechno.* 18: 3949.

[60] Regulacio, M.D. and M.Y. Han. 2010. Liquid crystal-material design and self-assembly. *Acc. Chem. Res.* 43: 621.

[61] Lan, G.Y., Y.W. Lin, Z.H. Lin and H.T. Chang. 2010. Solvothermal preparation of nano β HgS from a precursor. *J. Nanopart. Res.* 12: 1377.

[62] Zhang, J.B. and D.L. Zhang. 2010. Synthesis and growth kinetics of high quality InAs nanocrystals using in situ generated AsH 3 as the arsenic source. *Crystengcomm.* 12: 591.

[63] Petoud, S. 2009. Novel antenna for luminescent lanthanide cation emitting visible and near infrared—from smoll to poly metallic lanthanide containing nanocrystal. *Chimia* 63: 745.

[64] Tang, A.W., S.C. Qu, K. Li, Y.B. Hou, F. Teng, J. Cao, Y.S. Wang and Z.G. Wang. 2010. Controllable synthesis of silver and silver sulphide nanocrystal via selective cleavage of chemical bond. *Nanotechnology* 21: 285602.

[65] Wang, Y., Y.B. Hou, A.W. Tang, B. Feng, Y. Li and F. Teng. 2008. Liquid crystals. *Acta Phys-Chim Sin* 24: 296.

[66] Wang, F.D., R. Tang and W.E. Buhro. 2008. The trouble with TOPO; identification of adventitious impurities beneficial to the growth of cadmium selenide quantum dots, rods, and wires. *Nano Lett.* 8: 3521.

[67] Sudeep, P.K. and T. Emrick. 2009. Function Si and CdSe quantum dots-synthesis, conjugate formation and photoluminescence. *ACS Nano.* 3: 4105.

[68] Dollefeld, H., K. Hoppe, J. Kolny, K. Schilling, H. Weller and A. Eychmuller. 2002. Investigation on the stability of the thiol stabilized semiconductors nanoparticles. *Phys. Chem.* 4: 4747.

[69] Hwang, S., Y. Choi, S. Jeong, H. Jung, C.G. Kim, T.M. Chung and B.H. Ryu. 2010. Liquid crystal-material design and self-assembly. *Jpn. J. Appl. Phys.* 49: 05EA03.

[70] Rockenberger, J., L. Troger, A.L. Rogach, M. Tischer, M. Grundmann, A. Eychmuller and H. Weller. 1998. Thiol capped CdSe and CdTe nanocluster. *J. Chem. Phys.* 108: 7807.

[71] Talapin, D.V., S. Haubold, A.L. Rogach, A. Kornowski, M. Haase and H. Weller. 2001. A noval organometallic synthesis of highly luminescence nano-crystal. *J. Phys. Chem.* B 105: 2260.

[72] Gaponik, N., D.V. Talapin, A.L. Rogach, K. Hoppe, E.V. Shevchenko, A. Kornowski, A. Eychmuller and H. Weller. 2002. Thiol capped of CdTe nanocrystal. *J. Phys. Chem.* B 106: 7177.

[73] Rogach, A.L., L. Katsikas, A. Kornowski, D.S. Su, A. Eychmuller and H. Weller. 1996. Aqueous synthesis of CdTe nanocrystals. *Ber Bunsen Phys. Chem.* 100: 1772.

[74] Gaponik, N., D.V. Talapin, A.L. Rogach, A. Eychmuller and H. Weller. 2002. Thiol capped of CdTe nanocrystal. *Nano Lett.* 2: 803.

[75] Resch, U., A. Eychmuller, M. Haase and H. Weller. 1992. Absorption and fluorescence behaviour of redispersible CdS colloids in various organic solvent. *Langmuir* 8: 2215.

[76] Liu, I.S., H.H. Lo, C.T. Chien, Y.Y. Lin, C.W. Chen, Y.F. Chen, W.F. Su and S.C. Liou. 2008. Synthesis and characterization of CdSe nanocrystal in the presence of butylamine as a capping agent. *J. Mater. Chem.* 18: 675.

[77] Wang, X.S., T.E. Dykstra, M.R. Salvador, I. Manners, G.D. Scholes and M.A. Winnik. 2004. Surface passivation of luminescent colloidal quantum dot with poly(dimethylaninoethyl methcrylate) through a ligand exchange process. *J. Am. Chem. Soc.* 126: 7784.

[78] Foos, E.E., J. Wilkinson, A.J. Makinen, N.J. Watkins, Z.H. Kafafi and J.P. Long. 2006. Synthesis and surface composition study of CdSe nanoclusters prepared using solvent systems containing primary, secondary, and tertiary amines. *Chem. Mater.* 18: 2886.

[79] Zhou, L., C. Gao, X.Z. Hu and W.J. Xu. 2010. One-pot large-scale synthesis of robust ultrafine silica-hybridized CdTe quantum dots. *ACS Appl. Mater. Interfaces* 2: 1211.

[80] Li, H., W.Y. Shih and W.H. Shih. 2007. Stable aqeous ZnS quantum dots obtained using (3-mercaptopropyl) trimethoxy silane. *Nanotechnology* 18: 495605.

[81] Jana, N.R., C. Earhart and J.T. Ying. 2007. Synthesis of water and functionlized nanoparticals by silica coating. *Chem. Mater.* 19: 5074.

[82] Deuling, H.J., A. Buka and I. Janossy. 1976. Two freedericksz transition in crossed electric and magnetic field. *J. Phys-Paris* 37: 965.

[83] Freedericksz, V. and V. Zolina. 1933. Forces causing the orientation of an anisotropic liquid. *Trans Faraday Soc.* 29: 919.

[84] Brochard, F. and P.G.D. Gennes. 1970. Theory of magnetic suspension in liquid crystals. *J. Phys-Paris* 31: 691.

[85] Bilecka, I., I. Djerdj and M. Niederberger. 2008. One minutes synthesis of crystalline binary and ternary metal oxide nanoparticle. *Chem. Commun.* 886.

[86] Sun, S.H., H. Zeng, D.B. Robinson, S. Raoux, P.M. Rice, S.X. Wang and G.X. Li. 2004. Monodisperse MFe_2O_4(M = Fe, Co, Mn) nanoparticals. *J. Am. Chem. Soc.* 126: 273.

[87] Kim, E.H., H.S. Lee, B.K. Kwak and B.K. Kim. 2005. Synthesis of ferrofluid with magnetic nanoparticles by sonochemical method for MRI contrast agent. *J. Magn. Magn. Mater.* 289: 328.

[88] Massart, R. 1981. Preparation of aqueous magnetic liquid in alkaline and acidc media. *IEEE Trans Magn.* 17: 1247.

[89] Buchnev, E., A. Dyadyusha, M. Kaczmarek, V. Reshetnyak and Y. Reznikov. 2007. Enhanced two beam coupling in colloids of ferroelectric nanoparticles in liquid crystal. *J. Opt. Soc. Am.* B24: 1512.

[90] Buchnev, O., A. Glushchenko, Y. Reznikov, V. Reshetnyak, O. Tereshchenko and J. West. 2003. Nonlinear optics of liquid and photorefractive crystals. 5257: 7.

[91] Erdem, E., A. Matthes, R. Bottcher, H.J. Glasel and E. Hartmann. 2008. Size effect in ferroslectric $PbTiO_3$ nanoparticles observed by multi-frequency electron paramagnetic resonance spectroscopy. *J. Nanosci. Nanotechno.* 18: 702.

[92] Ram, S. and A. Mishra. 2006. A new ferroelectric PbZr0.52Ti0.48O3 polymorph of nanoparticles. *Mod. Phys Lett.* 20: 159.

[93] Ke, H., D.C. Jia, W. Wang and Y. Zhou. 2007. Liquid crystals: materials design and self-assembly. *Sol St. Phen.* 121–123: 843.

[94] Mornet, S., C. Elissalde, V. Hornebecq, O. Bidault, E. Duguet, A. Brisson and M. Maglione. 2005. Controlled growth of silica shell on $Ba_{0.6}To_3$ nanoparticles used As precursors of ferroelectric composites. *Chem. Mater.* 17: 4530.

[95] Cook, G., V.Y. Reshetnyak, R.F. Ziolo, S.A. Basun, P.P. Banerjee and D.R. Evans. 2010. Asymmetric freedericksz transition from symmetric liquid crystal cell doped with harvested ferroelectric nanoparticles. *Opt. Express* 18: 17339.

[96] Ram, S., A. Jana and T.K. Kundu. 2007. Ferroelectric $BaTiO_3$ phase of orthorhombic crystal structure contained nanoparticles. *J. Appl. Phys.* 102: 054107.

[97] Jha, A.K. and K. Prasad. 2010. Nanostructures for oral medicine. *Colloid Surface B* 75: 330.

[98] Bansal, V., P. Poddar, A. Ahmad and M. Sastry. 2006. Room temperature biosynthesis of ferroelectric barium titanate nanoparticle. *J. Am. Chem. Soc.* 128: 11958.

[99] Johann, F., T. Jungk, S. Lisinski, A. Hoffmann, L. Ratke and E. Soergel. 2010. Low voltage nanodomain writing in He-implanted lithium niobatecrystal. *Appl. Phys. Lett.* 96: 139901.

[100] Johann, F., T. Jungk, S. Lisinski, A. Hoffmann, L. Ratke and E. Soergel. 2009. Sol-gel derived ferroelectric nanoparticle investigated by piezoresponse force microscopy. *Appl. Phys. Lett.* 95: 202901.

[101] Tian, X.L., J.A. Li, K. Chen, J.A. Han, S.L. Pan, Y.J. Wang, X.Y. Fan, F. Li and Z.X. Zhou. 2010. Nearly monodisperse ferroelectric BaTiO3 hollow nanoparticles: Size-related solid evacuation in ostwald-ripening-induced hollowing process. *Cryst. Growth Des.* 10: 3990.

[102] Atkuri, H., G. Cook, D.R. Evans, C.I. Cheon, A. Glushchenko, V. Reshetnyak, Y. Reznikov, J. West and K. Zhang. 2009. Preparation of ferroelectric nanoparticle for their use in liquid crystalline colloids. *J. Opt. A-Pure Appl. Opt.* 11: 024006.

[103] Huang, X.H., S. Neretina and M.A. El-Sayed. 2009. Gold nanorods from synthesis and properties to biological and biomedical application. *Adv. Mater.* 21: 4880.

[104] Martin, C.R. 1996. Membrane based synthesis of nanoparticles. *Chem. Mater.* 8: 1739.

[105] Hurst, S.J., E.K. Payne, L.D. Qin and C.A. Mirkin. 2006. Multisegmented one dimension nanorods preparation by hard template synthetic method. *Angew Chem. Int. Ed.* 45: 2672.

[106] Chang, S.S., C.W. Shih, C.D. Chen, W.C. Lai and C.R.C. Wang. 1999. The shape transition of gold nanoparticle. *Langmuir* 15: 701.

[107] El-Sayed, M.A. 2001. Some interesting properties of metal confined in time and nanometer space of different shape. *Acc. Chem. Res.* 34: 257.

[108] Jana, N.R., L. Gearheart and C.J. Murphy. 2001. Wet chemical synthesis of high aspect ratio cylindrical gold nanorods. *Phys. Chem. B* 105: 4065.

[109] Murphy, C.J., T.K. San, A.M. Gole, C.J. Orendorff, J.X. Gao, L. Gou, S.E. Hunyadi and T. Li. 2005. Anisotropic metal nanoparticles: Synthesis, assembly and optical applications. *J. Phys. Chem. B* 109: 13857.

[110] Smith, D.K. and B.A. Korgel. 2008. The importance of the CTAB surfactant on the colloidal seed-mediated synthesis of gold nanorods. *Langmuir* 24: 644.

[111] Smith, D.K., N.R. Miller and B.A. Korgel. 2009. Iodide in CTAB prevents gold nanorod formation. *Langmuir* 25: 9518.

[112] Jana, N.R. 2005. Gram-scale synthesis of soluble, near-monodisperse gold nanorods and other anisotropic nanoparticles. *Small* 1: 875.

[113] Zijlstra, P., C. Bullen, J.W.M. Chon and M. Gu. 2006. High-temperature seedless synthesis of gold nanorods. *J. Phys. Chem. B* 110: 19315.

[114] Nikoobakht, B. and M.A. El-Sayed. 2003. Preparation and growth mechanism of gold nanorods (NRs) using seed-mediated growth method. *Chem. Mater.* 15: 1957.

[115] Iqbal, M., Y.I. Chung and G. Tae. 2007. An enhanced synthesis of gold nanorods by the addition of Pluronic (F-127) via a seed mediated growth process. *J. Mater. Chem.* 17: 335.

[116] Orendorff, C.J. and C.J. Murphy. 2006. Quantitation of metal content in the silver-assisted growth of gold nanorods. *J. Phys. Chem. B* 110: 3990.

[117] Liu, M.Z. and P. Guyot-Sionnest. 2005. Mechanism of silver(I)-assisted growth of gold nanorods and bipyramids. *J. Phys. Chem. B* 109: 22192.

[118] Perez-Juste, J., L.M. Liz-Marzan, S. Carnie, D.Y.C. Chan and P. Mulvaney. 2004. Electric-field-directed growth of gold nanorods in aqueous surfactant solutions. *Adv. Funct. Mater.* 14: 571.

[119] Sharma, V., K. Park and M. Srinivasarao. 2009. Colloidal dispersion of gold nanorods: Historical background, optical properties, seed-mediated synthesis, shape separation and self-assembly. *Mater. Sci. Eng. R Rep.* 65: 1.

[120] Park, H.J., A.H. CS, W.J. Kim, I.S. Choi, K.P. Lee and W.S. Yun. 2006. Temperature-induced control of aspect ratio of gold nanorods. *J. Vac. Sci. Technol. A* 24: 1323.

[121] Jana, N.R. 2003. Nanorod shape separation using surfactant assisted self-assembly. *Chem. Commun.* 1950.

[122] Mitamura, K. and T. Imae. 2009. Functionalization of gold nanorods toward their applications. *Plasmonics* 4: 23.

[123] Khoury, J.M., X.L.L. Zhou, L.T. Qu, L.M. Dai, A. Urbas and Q. Li. 2009. Liquid crystals: materials design and self-assembly. *Chem. Commun.* 2109.

[124] Mitamura, K., T. Imae, N. Saito and O. Takai. 2009. Liquid crystals: materials design and self-assembly. *Compos Interface* 16: 377.

[125] Wijaya, A. and K. Hamad-Schifferli. 2008. Ligand customization and DNA functionalization of gold nanorods via round-trip phase transfer ligand exchange. *Langmuir* 24: 9966.

[126] Pierrat, S., I. Zins, A. Breivogel and C. Sonnichsen. 2007. Self-assembly of small gold colloids with functionalized gold nanorods. *Nano Lett.* 7: 259.

[127] Wang, Z.L. 2003. Nanowires and Nanobelts: Materials, Properties and Devices. Kluwer, Dordrecht.

[128] Nalwa, H.S. 2000. Handbook of Nanostructured Materials and Nanotechnology. Academic Press, San Diego.

[129] Klabunde, K.J. 2001. Nanoscale Materials in Chemistry. Wiley, New York.

[130] Bockrath, M., D.H. Cobden, J. Lu, A.G. Rinzler, R.E. Smalley, L. Balents and P.L. McEuen. 1999. Luttinger-liquid behaviour in carbon nanotubes. *Nature* 397: 598.

[131] Yao, Z., H.W.C. Postma, L. Balents and C. Dekker. 1999. Carbon nanotube intramolecular junctions. *Nature* 402: 273.

[132] Wu, X.J., F. Zhu, C. Mu, Y.Q. Liang, L.F. Xu, Q.W. Chen, R.Z. Chen and D.S. Xu. 2010. Low-cost nanomaterials: toward greener and more efficient energy applications. *Coord. Chem. Rev.* 254: 1135.

[133] Xi, G. and J. Ye. 2010. Ultrathin SnO_2 nanorods: template- and surfactant-free solution phase synthesis, growth mechanism, optical, gas-sensing, and surface adsorption properties. *Inorg. Chem.* 49: 2302.

[134] Chun, J.Y. and J.W. Lee. 2010. Various synthetic methods for one-dimensional semiconductor nanowires/nanorods and their applications in photovoltaic devices. *Eur. J. Inorg. Chem.* 4251.

[135] Yi, G.C., C.R. Wang and W.I. Park. 2005. ZnO nanorods: synthesis, characterization and applications. *Semicond. Sci. Technol.* 20: S22.

[136] Lyu, S.C., Y. Zhang, C.J. Lee, H. Ruh and H.J. Lee. 2003. Low-temperature growth of ZnO nanowire array by a simple physical vapor-deposition method. *Chem. Mater.* 15: 3294.

[137] Ding, Y., P.X. Gao and Z.L. Wang. 2004. Catalyst–nanostructure interfacial lattice mismatch in determining the shape of VLS grown nanowires and nanobelts: a case of Sn/ZnO. *J. Am. Chem. Soc.* 126: 2066.

[138] Yu, J.H., J. Joo, H.M. Park, S.I. Baik, Y.W. Kim, S.C. Kim and T. Hyeon. 2005. Synthesis of quantum-sized cubic ZnS nanorods by the oriented attachment mechanism. *J. Am. Chem. Soc.* 127: 5662.

[139] Jana, N.R. and X.G. Peng. 2003. Single-phase and gram-scale routes toward nearly monodisperse Au and other noble metal nanocrystals. *J. Am. Chem. Soc.* 125: 14280.

[140] Park, J., K.J. An, Y.S. Hwang, J.G. Park, H.J. Noh, J.Y. Kim, J.H. Park, N.M. Hwang and T. Hyeon. 2004. Ultra-large-scale syntheses of monodisperse nanocrystals. *Nat. Mater.* 3: 891.

[141] Li, X., L.Y. Wang, L. Wang and Y.D. Li. 2008. Direct allylation of aldimines catalyzed by C 2-symmetric N,N'-dioxide–ScIII complexes: highly enantioselective synthesis of homoallylic amines. *Chem-Eur. J.* 14: 5951.

[142] Li, Y.D., H.W. Liao, Y. Ding, Y. Fan, Y. Zhang and Y.T. Qian. 1999. Solvothermal elemental direct reaction to CdE (E, S, Se, Te) semiconductor nanorod. *Inorg. Chem.* 38: 1382.

[143] Wu, X.J., F. Zhu, C. Mu, Y.Q. Liang, L.F. Xu, Q.W. Chen, R.Z. Chen and D.S. Xu. 2010. Electrochemical synthesis and applications of oriented and hierarchically quasi-1D semiconducting nanostructures. *Coord. Chem. Rev.* 254: 1135.

[144] Ajayan, P.M. 1999. Nanotubes from carbon. *Chem. Rev.* 99: 1787.

[145] Iijima, S. 1991. Helical microtubules of graphitic carbon. *Nature* 354: 56.

[146] Lu, W. and C.M. Lieber. 2007. Nanoelectronics from the bottom up. *Nat. Mater.* 6: 841.

[147] Peng, H. and X. Sun. 2009. Highly aligned carbon nanotube/polymer composite with much improved electrical conductivities. *Chem. Phys. Lett.* 471: 103.

[148] Gao, L., X. Zhou and Y. Ding. 2007. Effective thermal and electrical conductivity of carbon nanotube composites. *Chem. Phys. Lett.* 434: 297.

[149] Saito, R., G. Dresselhaus and M.S. Dresselhaus. 1998. Physical Properties of Carbon Nanotubes. Physical Properties of Carbon Nanotubes. Imperial College Press, London.

[150] Chae, H.G., J. Liu and S. Kumar. 2006. *Materials Carbon Nanotubes Properties and Applications.* Taylor & Francis Group, Boca Raton.

[151] Cinke, M., J. Li, B. Chen, A. Cassell, L. Delzeit, J. Han and M. Meyyappan. 2002. Selective gas detection using a carbon nanotube sensor. *Chem. Phys. Lett.* 365: 69.

[152] Ding, R.G., G.Q. Lu, Z.F. Yan and M.A. Wilson. 2001. Recent advances in the preparation and utilization of carbon nanotubes for hydrogen storage. *J. Nanosci. Nanotechnol.* 1: 7.

[153] Ko, H. and V.V. Tsukruk. 2006. Liquid-crystalline processing of highly oriented carbon nanotube arrays for thin-film transistors. *Nano Lett.* 6: 1443.

[154] Ahir, S.V. and E.M. Terentjev. 2005. Photomechanical actuation in polymer-nanotube composites. *Nat. Mater.* 4: 491.

[155] Zhang, M., S. Fang, A.A. Zakhidov, S.B. Lee, A.E. Aliev, C.D. Williams, K.R. Atkinson and R.H. Baughman. 2005. Strong, transparent, multifunctional, carbon nanotube sheets. *Science* 309: 1215.

[156] Du, F.M., J.E. Fischer and K.I. Winey. 2005. Effect of nanotube alignment on percolation conductivity in carbon nanotube/polymer composites. *Phys. Rev. B* 72: 121404.

[157] Scalia, G. 2010. Liquid crystal self organized soft functional materials for advanced application. *ChemPhysChem.* 11: 333.

[158] Itkis, M.E., D.E. Perea, R. Jung, S. Niyogi and R.C. Haddon. 2005. Comparison of analytical techniques for purity evaluation of single-walled carbon nanotubes. *J. Am. Chem. Soc.* 127: 3439.

[159] Murakami, Y., S. Chiashi, Y. Miyauchi, M.H. Hu, M. Ogura, T. Okubo and S. Maruyama. 2004. Growth of vertically aligned single-walled carbon nanotube films on quartz substrates and their optical anisotropy. *Chem. Phys. Lett.* 385: 298.

[160] Huang, S.M., X.Y. Cai, Chunsheng Du and J. Liu. 2003. Oriented long single walled carbon nanotubes on substrates from floating catalysts. *J. Am. Chem. Soc.* 125: 5636.

[161] Zhang, S.J, Q.W. Li, I.A. Kinloch and A.H. Windle. 2010. Ordering in a droplet of an aqueous suspension of single-wall carbon nanotubes on a solid substrate. *Langmuir* 26: 2107.

[162] Lee, H.W., W. You, S. Barman, S. Hellstrom, M.C. LeMieux, J.H. Oh, S. Liu, T. Fujiwara, W.M. Wang, B. Chen, Y.W. Jin, J.M. Kim and Z.A. Bao. 2009. Lyotropic liquid-crystalline solutions of high-concentration dispersions of single-walled carbon nanotubes with conjugated polymers. *Small* 5: 1019.

[163] Davis, V.A., L.M. Ericson, A.N.G. Parra-Vasquez, H. Fan, Y.H. Wang, V. Prieto, J.A. Longoria, S. Ramesh, R.K. Saini, C. Kittrell, W.E. Billups, W.W. Adams, R.H. Hauge, R.E. Smalley and M. Pasquali. 2004. Phase behavior and rheology of SWNTs in superacids. *Macromolecules* 37: 154.

[164] Green, M.J., A.N.G. Parra-Vasquez, N. Behabtu and M. Pasquali. 2009. Modeling the phase behavior of polydisperse rigid rods with attractive interactions with applications to single-walled carbon nanotubes in superacids. *J. Chem. Phys.* 131: 084901.

[165] Badaire, S., C. Zakri, M. Maugey, A. Derre, J.N. Barisci, G. Wallace and P. Poulin. 2005. Liquid crystals of DNA-stabilized carbon nanotubes. *Adv. Mater.* 17: 1673.

[166] Lu, L. and W. Chen. 2010. Large-scale aligned carbon nanotubes from their purified, highly concentrated suspension. *ACS Nano* 4: 1042.

[167] Vijayakumar, V.N. and M.M.L.N. Madhu. 2010. Dispersion of multi walled carbon nanotubes in a hydrogen bonded liquid crystal. *Phys B* 405: 4418.

[168] Bravo-Sanchez, M., T.J. Simmons and M.A. Vidal. 2010. Liquid crystal behavior of single wall carbon nanotubes. *Carbon* 48: 3531.

[169] Puech, N., E. Grelet, P. Poulin, C. Blanc and P. van der Schoot. 2010. Nematic droplets in aqueous dispersions of carbon nanotube. *Phys. Rev. E* 82: 020702.

[170] Samson, J., A. Varotto, P.C. Nahirney, A. Toschi, I. Piscopo and C.M. Drain. 2009. Fabrication of metal nanoparticles using toroidal plasmid DNA as a sacrificial mold. *ACS Nano* 3: 339.

[171] Yan, H.W., J.B. Hou, Z.P. Fu, B.F. Yang, P.H. Yang, K.P. Liu, M.W. Wen, Y.J. Chen, S.Q. Fu and F.Q. Li. 2009. Growth and photocatalytic properties of one-dimensional ZnO nanostructures prepared by thermal evaporation. *Mater. Res. Bull.* 44: 1954.

[172] Polleux, J., N. Pinna, M. Antonietti and M. Niederberger. 2005. Growth and assembly of crystalline tungsten oxide nanostructures assisted by bioligation. *J. Am. Chem. Soc.* 127: 15595.

[173] Ezawa, M. 2009. Quasi-ferromagnet spintronics in the graphene nanodisc–lead system. *New J. Phys.* 11: 095005.

[174] Sigman, M.B., A. Ghezelbash, T. Hanrath, A.E. Saunders, F. Lee and B.A. Korgel. 2003. Solventless synthesis of monodisperse Cu2S nanorods, nanodisks, and nanoplatelets. *J. Am. Chem. Soc.* 125: 16050.

[175] Saunders, A.E., A. Ghezelbash, D.M. Smilgies, M.B. Sigman and B.A. Korgel. 2006. Columnar self-assembly of colloidal nanodisks. *Nano Lett.* 6: 2959.

[176] Brown, A.B.D., S.M. Clarke and A.R. Rennie. 1998. Ordered phase of platelike particles in concentrated dispersions. *Langmuir* 14: 3129.

[177] van der Kooij, F.M., K. Kassapidou and H.N.W. Lekkerkerker. 2000. Liquid crystal phase transitions in suspensions of polydisperse plate-like particles. *Nature* 406: 868.

[178] Davidson, P. and J.C.P. Gabriel. 2005. Mineral liquid crystals. *Curr. Opin. Colloid Interface Sci.* 9: 377.

[179] Ok, C.H., B.Y. Kim, B.Y. Oh, Y.H. Kim, K.M. Lee, H.G. Park, J.M. Han, D.S. Seo, D.K. Lee and J.Y. Hwang. 2008. IPS mode investigation of liquid crystal alignment on organic hybrid overcoat layer via ion beam irradiation. *LiqCryst.* 35: 1373.

[180] Gwag, J.S. 2010. In-plane switching liquid crystal mode for high brightness. *Opt. Express* 16: 12220.

[181] Hwang, J.Y., S.H. Kim, S.H. Choi, H.K. Kang, J.H. Choi, M.H. Ham, J.M. Myoung and D.S. Seo. 2006. EO characteristics of fringe-field switching LCD on a-C: H thin films using the UV alignment method. *Ferroelectrics* 344: 435.

[182] Shin, H.K., K.H. Kim, T.H. Yoon and J.C. Kim. 2008. Vertical alignment nematic liquid crystal cell controlled by double-side in-plane switching with positive dielectric anisotropy liquid crystal. *J. Appl. Phys.* 104: 084515.

[183] Lu, R.B., X.Y. Nie and S.T. Wu. 2008. Color performance of an MVA-LCD using an LED backlight. *J. Soc. Inf. Disp.* 16: 1139.

[184] Voloschenko, D., O.P. Pishnyak, S.V. Shiyanovskii and O.D. Lavrentovich. 2002. Effect of director distortions on morphologies of phase separation in liquid crystal. *Phys. Rev. E* 65: 060701.

[185] Nishida, N., Y. Shiraishi, S. Kobayashi and N. Toshima. 2008. Fabrication of liquid crystal sol containing capped Ag–Pd bimetallic nanoparticles and their electro-optic properties. *J. Phys. Chem. C* 112: 20284.

[186] Khatua, S., P. Manna, W.S. Chang, A. Tcherniak, E. Friedlander, E.R. Zubarev and S. Link. 2010. Plasmonic nanoparticles–liquid crystal composites. *J. Phys. Chem. C* 114: 7251.

[187] Yoshida, H., K. Kawamoto, H. Kubo, T. Tsuda, A. Fujii, S. Kuwabata and M. Ozaki. 2010. Nanoparticle-dispersed liquid crystals fabricated by sputter doping. *Adv. Mater.* 22: 622.

[188] Blach, J.F., S. Saitzek, C. Legrand, L. Dupont, J.F. Henninot and M. Warenghem. 2010. BaTiO$_3$ ferroelectric nanoparticles dispersed in 5CB nematic liquid crystal: Synthesis and electro-optical characterization. *J. Appl. Phys.* 107: 074102.

[189] Glushchenko, A., C.I. Cheon, J. West, F.H. Li, E. Buyuktanir, Y. Reznikov and A. Buchnev. 2006. Liquid crystals: materials design and self-assembly. *Mol. CrystLiqCryst.* 453: 227.

[190] Ouskova, E., O. Buchnev, V. Reshetnyak, Y. Reznikov and H. Kresse. 2003. Dielectric relaxation spectroscopy of a nematic liquid crystal doped with ferroelectric Sn$_2$P$_2$S$_6$ nanoparticles. *LiqCryst.* 30: 1235.

[191] West, J.L., F.H. Li, K. Zhang and H. Atkuri. 2007. AD'07: Proc Asia Display 2007, Liquid Crystals: Materials Design and Self-assembly. vols. 1 and 2, p 113.

[192] Reshetnyak, V.Y., S.M. Shelestiuk and T.J. Sluckin. 2006. Fredericksz transition threshold in nematic liquid crystals filled with ferroelectric nano-particles. *Mol. Cryst. Liq. Cryst.* 454: 201.

[193] Reznikov, Y., O. Buchnev, A. Glushchenko, V. Reshetnyak, O. Tereshchenko and J. West. 2005. Ferroelectric particles-liquid crystal dispersions. Proc SPIE - Emerging Liquid Crystal Technologies 5741: 171.

[194] Reznikov, Y., O. Buchnev, O. Tereshchenko, V. Reshetnyak, A. Glushchenko and J. West. 2003. Ferroelectric nematic suspension. *Appl. Phys. Lett.* 82: 1917.

[195] Li, F.H., O. Buchnev, C.I. Cheon, A. Glushchenko, V. Reshetnyak, Y. Reznikov, T.J. Sluckin and J.L. West. 2006. Orientational coupling amplification in ferroelectric nematic colloids. *Phys. Rev. Lett.* 97: 147801.

[196] Glushchenko, A., Il F. C. Cheon, J. West and Y. Reznikov. 2007. Proc SPIE – Emerging Liquid Crystal Technologies II 6487: T4870.

[197] Li, F.H., J. West, A. Glushchenko, C.I. Cheon and Y. Reznikov. 2006. Liquid crystals: materials design and self-assembly. *J. Soc. Inf. Disp.* 14: 523.

[198] Li, F.H., J. West, A. Glushchenko, C.I. Cheon and Y. Reznikov. 2006. Ferroelectric nanoparticle/liquid-crystal colloids for display applications. *J. Soc. Inf. Disp.* 14: 523.

[199] Tian, P., G.D. Smith and M. Glaser. 2006. Molecular dynamics simulations studies of nanoparticles in an isotropic liquid crystal matrix: Single particle behavior and pairwise interactions. *J. Chem. Phys.* 124: 161101.

[200] Xu, J.Q., D. Bedrov, G.D. Smith and M.A. Glaser. 2009. Molecular dynamics simulation study of spherical nanoparticles in a nematogenic matrix: Anchoring, interactions, and phase behaviour. *Phys. Rev. E* 79: 011704.

[201] Piegdon, K.A., S. Declair, J. Forstner, T. Meier, H. Matthias, M. Urbanski, H.S. Kitzerow, D. Reuter, A.D. Wieck, A. Lorke and C. Meier. 2010. Tuning quantum-dot based photonic devices with liquid crystals. *Opt. Express* 18: 7946.

[202] Tong, X. and Y. Zhao. 2007. Liquid-crystal gel-dispersed quantum dots: reversible modulation of photoluminescence intensity using an electric field. *J. Am. Chem. Soc.* 129: 6372.

[203] Zhang, T., C. Zhong and J. Xu. 2009. CdS-nanoparticle-doped liquid crystal displays showing low threshold voltage. *Jpn. J. Appl. Phys.* 48: 055002.

[204] Chen, W.T., P.S. Chen and C.Y. Chao. 2009. Effect of doped insulating nanoparticles on the electro-optical characteristics of nematic liquid crystals. *Jpn. J. Appl. Phys.* 48: 015006.

[205] Manohar, R., S.P. Yadav, A.K. Srivastava, A.K. Misra, K.K. Pandey, P.K. Sharma and A.C. Pandey. 2009. Zinc oxide (1% Cu) nanoparticle in nematic liquid crystal: dielectric and electro-optical study. *Jpn. J. Appl. Phys.* 48: 101501.

[206] Hwang, S.J., S.C. Jeng, C.Y. Yang, C.W. Kuo and C.C. Liao. 2009. Characteristics of nanoparticle-doped homeotropic liquid crystal devices. *J. Phys. D Appl. Phys.* 42: 025102.

[207] Kuo, C.W., S.C. Jeng, H.L. Wang and C.C. Liao. 2007. Application of nanoparticle-induced vertical alignment in hybrid-aligned nematic liquid crystal cell. *Appl. Phys. Lett.* 91: 141103.

[208] Masutani, A., T. Roberts, B. Schuller, N. Hollfelder, P. Kilickiran, A. Sakaigawa, G. Nelles and A. Yasuda. 2008. Liquid crystals: materials design and self-assembly. *J. Soc. Inf. Disp.* 16: 137.

[209] Huang, C.Y., C.C. Lai, Y.H. Tseng, T. Yang, C.J. Tien and K.Y. Lo. 2008. Silica-nanoparticle-doped nematic display with multistable and dynamic modes. *Appl. Phys. Lett.* 92: 221908.

[210] Hu, W., H.Y. Zhao, L.K. Shan, L. Song, H. Cao, Z. Yang, Z.H. Cheng, C.Z. Yan, S.J. Li, H.A. Yang and L. Guo. 2010. Liquid crystals: materials design and self-assembly. *LiqCryst.* 37: 563.

[211] Qi, H. and T. Hegmann. 2006. Liquid crystals materials design and self-assembly. *J. Mater. Chem.* 16: 4197.

[212] Pratibha, R., W. Park and I.I. Smalyukh. 2010. Colloidal gold nanosphere dispersions in smectic liquid crystals and thin nanoparticle-decorated smectic films. *J. Appl. Phys.* 107: 063511.

[213] Cordoyiannis, G., S. Kralj, G. Nounesis, Z. Kutnjak and S. Zumer. 2007. Liquid crystals: materials design and self-assembly. *Phys. Rev. E* 75: 021702.

[214] Martinez-Miranda, L.J., K. McCarthy, L.K. Kurihara, J.J. Harry and A. Noel. 2006. Effect of the surface coating on the magnetic nanoparticle smectic-A liquid crystal interaction. *Appl. Phys. Lett.* 89: 161917.

[215] Heppke, G., D. Kruerke, C. Lohning, D. Lotzsch, D. Moro, M. Muller and H. Sawade. 2000. Liquid crystals: materials design and self-assembly. *J. Mater. Chem.* 10: 2657.

[216] Seshadri, T. and H.J. Haupt. 1998. Liquid crystals: materials design and self-assembly. *Chem. Commun.* 735.

[217] Stegemeyer, H., M. Schumacher and E. Demikhov. 1993. Liquid crystals: materials design and self-assembly. *LiqCryst.* 15: 933.

[218] Kitzerow, H.S., H. Schmid, A. Ranft, G. Heppke, R.A.M. Hikmet and J. Lub. 1993. Liquid crystals: materials design and self-assembly. *LiqCryst.* 14: 911.

[219] Stegemeyer, H., H. Onusseit and H. Finkelmann. 1989. Formation of a blue phase in a liquid-crystalline side chain polysiloxane Makromol. *Chem-Rapid* 10: 571.

[220] Chanishvili, A.G., G.S. Chilaya, Z.M. Elashvili, S.P. Ivchenko, D.G. Khoshtaria and K.D. Vinokur. 1986. Liquid crystals: materials design and self-assembly. *Mol. CrystLiqCryst.* 3: 91.

[221] Marcus, M.A. 1984. Liquid crystals: materials design and self-assembly. *Mol. CrystLiqCryst.* 102: 207.

[222] Kitzerow, H.S., P.P. Crooker, J. Rand, J. Xu and G. Heppke. 1992. Liquid crystals: materials design and self-assembly. *J. Phys. II* 2: 279.

[223] Stark, H. and H.R. Trebin. 1991. Theory of electrostriction of liquid-crystalline blue phases I and II. *Phys. Rev. A* 44: 2752.

[224] Yang, D.K. and P.P. Crooker. 1990. Polymer-dispersed chiral liquid crystal color display. *LiqCryst.* 7: 411.

[225] Kitzerow, H.S., P.P. Crooker, S.L. Kwok, J. Xu and G. Heppke. 1990. Dynamics of blue-phase selective reflections in an electric field. *Phys. Rev. A* 42: 3442.

[226] Kitzerow, H.S., P.P. Crooker, S.L. Kwok and G. Heppke. 1990. Liquid crystals: materials design and self-assembly. *J. Phys-Paris* 51: 1303.

[227] Porsch, F. and H. Stegemeyer. 1989. Liquid crystals: materials design and self-assembly. *LiqCryst.* 5: 791.

[228] Heppke, G., B. Jerome, H.S. Kitzerow and P. Pieranski. 1989. Electrostriction of the cholesteric blue phases BPI and BPII in mixtures with positive dielectric anisotropy. *J. Phys-Paris* 50: 2991.

[229] Heppke, G., B. Jerome, H.S. Kitzerow and P. Pieranski. 1989. Liquid crystals: materials design and self-assembly. *J. Phys-Paris* 50: 549.

[230] Pieranski, P., P.E. Cladis and R. Barbetmassin. 1989. Liquid crystals: materials design and self-assembly. *LiqCryst.* 5: 829.

[231] Pieranski, P., R. Barbetmassin and P.E. Cladis. 1985. Electrostriction of BPI and BPII for blue phase systems with negative dielectric anisotropy. *Phys. Rev. A* 31: 3912.

[232] Cao, W.Y., A. Munoz, P. Palffy-Muhoray and B. Taheri. 2002. Lasing in a three-dimensional photonic crystal of the liquid crystal blue phase II. *Nat. Mater.* 1: 111.

[233] Kitzerow, H.S. 2006. Liquid crystals: materials design and self-assembly. *Chem. Phys. Chem.* 7: 63.

[234] Kitzerow, H.S. 2010. Blue phases: prior art, potential polar effects, challenges. *Ferroelectrics* 395: 66.

[235] Coles, H.J. and M.M.N. Pivnenko. 2005. Liquid crystal 'blue phases' with a wide temperature range. *Nature* 436: 997.

[236] Kikuchi, H., M. Yokota, Y. Hisakado, H. Yang and T. Kajiyama. 2002. Polymer-stabilized liquid crystal blue phases. *Nat. Mater.* 1: 64.

[237] Rao, L.H., Z.B. Ge and S.T. Wu. 2010. Viewing angle controllable displays with a blue-phase liquid crystal cell. *Opt. Express* 18: 3143.

[238] Rao, L.H., H.C. Cheng and S.T. Wu. 2010. Low voltage blue-phase LCDs with double-penetrating fringe fields. *J. DispTechnol.* 6: 287.

[239] Lu, S.Y. and L.C. Chien. 2010. Electrically switched color with polymer-stabilized blue-phase liquid crystals. *Opt. Lett.* 35: 562.

[240] Yokoyama, S., S. Mashiko, H. Kikuchi, K. Uchida and T. Nagamura. 2006. Laser emission from a polymer-stabilized liquid-crystalline blue phase. *Adv. Mater.* 18: 48.

[241] Morris, S.M., A.D. Ford, C. Gillespie, M.N. Pivnenko, O. Hadeler and H.J. Coles. 2006. Liquid crystals: materials design and self-assembly. *J. Soc. Inf. Disp.* 14: 565.

[242] Liu, H.Y., C.T. Wang, C.Y. Hsu, T.H. Lin and J.H. Liu. 2010. Optically tuneable blue phase photonic band gaps. *Appl. Phys. Lett.* 96: 121103.

[243] Coles, H.J. 2007. Proc SPIE - Emerging Liquid Crystal Technologies II 6487: M4870.

[244] Yoshida, H., Y. Tanaka, K. Kawamoto, H. Kubo, T. Tsuda, A. Fujii, S. Kuwabata, H. Kikuchi and M. Ozaki. 2009. Nanoparticle-stabilized cholesteric blue phases. *Appl. Phys. Express* 2: 121501.

[245] Zhou, X.L., S.W.Kang, S.Kumar, Q.Li, (2009). Liquid Crystals: Materials Design and Self-assembly. *LiqCryst* 36:269

[246] Li, L.F., S.W. Kang, J. Harden, Q.J. Sun, X.L. Zhou, L.M. Dai, A. Jakli, S. Kumar and Q. Li. 2008. Nature-inspired light-harvesting liquid crystalline porphyrins for organic photovoltaics. *LiqCryst.* 35: 233.

[247] Kumar, S., S.K. Pal, P.S. Kumar and V. Lakshminarayanan. 2007. Novel conducting nanocomposites: synthesis of triphenylene-covered gold nanoparticles and their insertion into a columnar matrix. *Soft Matter* 3: 896.

[248] Kumar, S. 2007. Liquid crystals: materials design and self-assembly. *Synth. React. Inorg.* Me 37: 327.

[249] Holt, L.A., R.J. Bushby, S.D. Evans, A. Burgess and G. Seeley. 2008. A 10^6-fold enhancement in the conductivity of a discotic liquid crystal doped with only 1% (w/w) gold nanoparticles. *J. Appl. Phys.* 103: 063712.

[250] Durr, N.J., T. Larson, D.K. Smith, B.A. Korgel, K. Sokolov and A. Ben-Yakar. 2007. Two-photon luminescence imaging of cancer cells using molecularly targeted gold nanorods. *Nano Lett.* 7: 941.

[251] Perez-Juste, J., B. Rodriguez-Gonzalez, P. Mulvaney and L.M. Liz-Marzan. 2005. Optical control and patterning of gold-nanorod–poly(vinyl alcohol) nanocomposite films. *Adv. Funct. Mater.* 15: 1065.

[252] Wilson, O., G.J. Wilson and P. Mulvaney. 2002. Laser writing in polarized silver nanorod films. *Adv. Mater.* 14: 1000.

[253] Zhang, S.S., G. Leem, L.O. Srisombat and T.R. Lee. 2008. Rationally designed ligands that inhibit the aggregation of large gold nanoparticles in solution. *J. Am. Chem. Soc.* 130: 113.

[254] Liu, Q.K., Y.X. Cui, D. Gardner, X. Li, S.L. He and I.I. Smalyukh. 2010. Self-alignment of plasmonic gold nanorods in reconfigurable anisotropic fluids for tunable bulk metamaterial applications. *Nano Lett.* 10: 1347.

[255] Sridevi, S., S.K. Prasad, G.G. Nair, V. D'Britto and B.L.V. Prasad. 2010. Enhancement of anisotropic conductivity, elastic, and dielectric constants in a liquid crystal-gold nanorod system. *Appl. Phys. Lett.* 97: 151913.

[256] Evans, P.R., G.A. Wurtz, W.R. Hendren, R. Atkinson, W. Dickson, A.V. Zayats and R.J. Pollard. 2007. Electrically switchable nonreciprocal transmission of plasmonic nanorods with liquid crystal. *Appl. Phys. Lett.* 91: 043101.

[257] Chu, K.C., C.Y. Chao, Y.F. Chen, Y.C. Wu and C.C. Chen. 2006. Electrically controlled surface plasmon resonance frequency of gold nanorods. *Appl. Phys. Lett.* 89: 103107.

[258] Lapointe, C.P., D.H. Reich and R.L. Leheny. 2008. Manipulation and organization of ferromagnetic nanowires by patterned nematic liquid crystals. *Langmuir* 24: 11175.

[259] Chen, H.S., C.W. Chen, C.H. Wang, F.C. Chu, C.Y. Chao, C.C. Kang, P.T. Chou and Y.F. Chen. 2010. Color-tunable light-emitting device based on the mixture of CdSe nanorods and dots embedded in liquid-crystal cells. *J. Phys. Chem. C* 114: 7995.

[260] Wu, K.J., K.C. Chu, C.Y. Chao, Y.F. Chen, C.W. Lai, C.C. Kang, C.Y. Chen and P.T. Chou. 2007. CdS nanorods imbedded in liquid crystal cells for smart optoelectronic devices. *Nano Lett.* 7: 1908.

[261] Acharya, S., S. Kundu, J.P. Hill, G.J. Richards and K. Ariga. 2009. Liquid crystals: materials design and self-assembly. *Adv. Mater.* 21: 989.

[262] Kundu, S., J.P. Hill, G.J. Richards, K. Ariga, A.H. Khan, U. Thupakula and S. Acharya. 2010. Ultranarrow PbS nanorod-nematic liquid crystal blend for enhanced electro-optic properties. *ACS Appl. Mater. Interfaces* 2: 2759.

[263] Ouskova, E., O. Buluy, C. Blanc, H. Dietsch and A. Mertelj. 2010. Enhanced magneto-optical properties of suspensions of spindle type mono-dispersed hematite nano-particles in liquid crystal. *Mol. CrystLiqCryst.* 104.

[264] Williams, Y., K. Chan, J.H. Park, I.C. Khoo, B. Lewis and T.E. Mallouk. 2005. Electro-optical and nonlinear optical properties of semiconductor nanorod doped liquid crystals. Proc SPIE – Liquid Crystals IX 5936: 593613.

[265] Zorn, M., S. Meuer, M.N. Tahir, Y. Khalavka, C. Sonnichsen, W. Tremel and R. Zentel. 2008. Liquid crystalline phases from polymer functionalised semiconducting nanorods. *J. Mater. Chem.* 18: 3050.

[266] Meuer, S., K. Fischer, I. Mey, A. Janshoff, M. Schmidt and R. Zentel. 2008. Liquid crystals from polymer-functionalized TiO$_2$ nanorod mesogens. *Macromolecules* 41: 7946.

[267] Zorn, M. and R. Zentel. 2008. Liquid crystalline orientation of semiconducting nanorods in a semiconducting matrix. *Macromol. Rapid Commun.* 29: 922.

[268] Zorn, M., M.N. Tahir, B. Bergmann, W. Tremel, C. Grigoriadis, G. Floudas and R. Zentel. 2010. Functional polymers by post-polymerization modification. *Macromol. Rapid Commun.* 31: 1101.

[269] Popa-Nita, V. and S. Kralj. 2010. Liquid crystal-carbon nanotubes mixtures. *J. Chem. Phys.* 132: 024902.

[270] Scalia, G., J.P.F. Lagerwall, S. Schymura, M. Haluska, F. Giesselmann and S. Roth. 2007. Carbon nanotubes in liquid crystals as versatile functional materials. *Phys. Status Solidi B* 244: 4212.

[271] Lagerwall, J.P.F. and G. Scalia. 2008. Carbon nanotubes in liquid crystals. *J. Mater. Chem.* 18: 2890.

[272] Dierking, I., G. Scalia and P. Morales. 2005. Liquid crystal–carbon nanotube dispersions. *J. Appl. Phys.* 97: 044309.

[273] Dierking, I., G. Scalia, P. Morales and D. LeClere. 2004. Aligning and reorienting carbon nanotubes with nematic liquid crystals. *Adv. Mater.* 16: 865.

[274] Schymura, S., M. Kuehnast, V. Lutz, S. Jagiella, U. Dettlaff-Weglikowska, S. Roth, F. Giesselmann, C. Tschierske, G. Scalia and J. Lagerwall. 2010. Towards efficient dispersion of carbon nanotubes in thermotropic liquid crystals. *Adv. Funct. Mater.* 20: 3350.

[275] Kuehnast, M., C. Tschierske and J. Lagerwall. 2010. Tailor-designed polyphilic promotors for stabilizing dispersions of carbon nanotubes in liquid crystals. *Chem. Commun.* 46: 6989.

[276] Kimura, M., N. Miki, N. Adachi, Y. Tatewaki, K. Ohta and H. Shirai. 2009. Organization of single-walled carbon nanotubes wrapped with liquid-crystalline ϖ-conjugated oligomers. *J. Mater. Chem.* 19: 1086.

[277] Trushkevych, O., N. Collings, T. Hasan, V. Scardaci, A.C. Ferrari, T.D. Wilkinson, W.A. Crossland, W.I. Milne, J. Geng, B.F.G. Johnson and S. Macaulay. 2008. Characterization of carbon nanotube–thermotropic nematic liquid crystal composites. *J. Appl. Phys.* 41: 125106.

[278] Jeon, S.Y., K.A. Park, I.S. Baik, S.J. Jeong, S.H. Jeong, K.H. An, S.H. Lee and Y.H. Lee. 2007. Liquid crystals: materials design and self-assembly. *Nano* 2: 41.

[279] Moore, V.C., M.S. Strano, E.H. Haroz, R.H. Hauge, R.E. Smalley, J. Schmidt and Y. Talmon. 2003. Individually suspended single-walled carbon nanotubes in various surfactants. *Nano Lett.* 3: 1379.

[280] Heister, E., C. Lamprecht, V. Neves, C. Tilmaciu, L. Datas, E. Flahaut, B. Soula, P. Hinterdorfer, H.M. Coley, S.R.P. Silva and J. McFadden. 2010. Higher dispersion efficacy of functionalized carbon nanotubes in chemical and biological environments. *ACS Nano* 4: 2615.

[281] Lagerwall, J.P.F., G. Scalia, M. Haluska, U. Dettlaff-Weglikowska, F. Giesselmann and S. Roth. 2006. Simultaneous alignment and dispersion of carbon nanotubes with lyotropic liquid crystals. *Phys. Status Solidi B* 243: 3046.

[282] Lagerwall, J., G. Scalia, M. Haluska, U. Dettlaff-Weglikowska, S. Roth and F. Giesselmann. 2007. Nanotube alignment using lyotropic liquid crystals. *Adv. Mater.* 19: 359.

[283] Weiss, V., R. Thiruvengadathan and O. Regev. 2006. Preparation and characterization of a carbon nanotube–lyotropic liquid crystal composite. *Langmuir* 22: 854.

[284] Scalia, G., C. von Buhler, C. Hagele, S. Roth, F. Giesselmann and J.R.F. Lagerwall. 2008. Spontaneous macroscopic carbon nanotube alignment via colloidal suspension in hexagonal columnar lyotropic liquid crystals. *Soft Matter* 4: 570.

[285] Schymura, S., E. Enz, S. Roth, G. Scalia and J.R.F. Lagerwall. 2009. Macroscopic-scale carbon nanotube alignment via self-assembly in lyotropic liquid crystals. *Synth Metals* 159: 2177.

[286] Goncharuk, A.I., N.I. Lebovka, L.N. Lisetski and S.S. Minenko. 2009. Aggregation, percolation and phase transitions in nematic liquid crystal EBBA doped with carbon nanotubes. *J. Phys. D: Appl. Phys.* 42: 165411.

[287] Chen, H.Y., W. Lee and N.A. Clark. 2007. Faster electro-optical response characteristics of a carbon-nanotube-nematic suspension. *Appl. Phys. Lett.* 90: 033510.

[288] Jeon, S.Y., S.H. Shin, S.J. Jeong, S.H. Lee, S.H. Jeong, Y.H. Lee, H.C. Choi and K.J. Kim. 2007. Liquid crystals: materials design and self-assembly. *Appl. Phys. Lett.* 90: 121901.

[289] Lisetski, L.N., S.S. Minenko, A.P. Fedoryako and N.I. Lebovka. 2009. Optical transmission of nematic liquid crystal 5CB doped by single-walled and multi-walled carbon nanotubes. *Phys E* 41: 431.

[290] Zhao, W., J. Wang, J. He, L. Zhang, X. Wang and R. Li. 2009. Preparation and characterization of carbon nanotubes/monotropic liquid crystal composites. *Appl. Surf. Sci.* 255: 6589.

[291] Rahman, M. and W. Lee. 2009. Scientific duo of carbon nanotubes and nematic liquid crystals. *J. Phys. D: Appl. Phys.* 42: 063001.

[292] Kovalchuk, A., L. Dolgov and O. Yaroshchuk. 2008. Liquid crystals: materials design and self-assembly. *Semicond. Phys, Quantum Electron Optoelectron* 11: 337.

[293] Basu, R., K.P. Sigdel and G.S. Iannacchione. 2009. Isotropic to nematic phase transition in carbon nanotube dispersed liquid crystal composites. arXiv. org, e-Print Arch., Condens. Matter, p 1.

[294] Arora, P., A. Mikulko, F. Podgornov and W. Haase. 2009. Complementary studies of BaTiO$_3$ nanoparticles suspended in a ferroelectric liquid-crystalline mixture. *Mol. CrystLiqCryst.* 502: 1.
[295] Podgornov, F.V., A.M. Suvorova, A.V. Lapanik and W. Haase. 2009. Electrooptic and dielectric properties of ferroelectric liquid crystal/single walled carbon nanotubes dispersions confined in thin cells. *Chem. Phys. Lett.* 479: 206.
[296] Lagerwall, J.P.F., R. Dabrowski and G. Scalia. 2007. Carbon nanoparticles in cholesteric liquid crystals. *J. Non-Cryst. Solids* 353: 4411.
[297] Lisetski, L.N., S.S. Minenko, A.V. Zhukov, P.P. Shtifanyuk and N.I. Lebovka. 2009. Dispersions of carbon nanotubes in cholesteric liquid crystals. *Mol. CrystLiqCryst.* 510: 43.
[298] Pizzey, C., J. Van Duijneveldt and S. Klein. 2004. Liquid crystals: materials design and self-assembly. *Mol. CrystLiqCryst.* 409: 51.
[299] Pizzey, C., S. Klein, E. Leach, J.S. van Duijneveldt and R.M. Richardson. 2004. Suspensions of colloidal plates in a nematic liquid crystal: a small angle X-ray scattering study. *J. Phys-Condens. Mat.* 16: 2479.
[300] van Duijneveldt, J.S., S. Klein, E. Leach, C. Pizzey and R.M. Richardson. 2005. Large scale structures in liquid crystal/clay colloids. *J. Phys. CondensMatter* 17: 2255.
[301] Zhang, Z.X. and van Duijneveldt, J.S. 2007. Effect of suspended clay particles on isotropic–nematic phase transition of liquid crystal. *Soft Matter* 3: 596.
[302] Bezrodna, T., I. Chashechnikova, V. Nesprava, G. Puchkovska, Y. Shaydyuk, Y. Boyko, J. Baran and M. Drozd. 2010. Structure and spectroscopic properties of organoclays doped by multiwall carbon nanotubes. *Liq. Cryst.* 37: 263.
[303] Chashechnikova, I., L. Dolgov, T. Gavrilko, G. Puchkovska, Y. Shaydyuk, N. Lebovka, V. Moraru, J. Baran and H. Ratajczak. 2005. Optical properties of heterogeneous nanosystems based on montmorillonite clay mineral and 5CB nematic liquid crystal. *J. Mol. Struct.* 744: 563.
[304] Bezrodna, T., I. Chashechnikova, L. Dolgov, G. Puchkovska, Y. Shaydyuk, N. Lebovka, V. Moraru, J. Baran and H. Ratajczak. 2005. Effects of montmorillonite modification on optical properties of heterogeneous nematic liquid crystal–clay mineral nanocomposites. *LiqCryst.* 32: 1005.
[305] Chang, Y.M., T.Y. Tsai, Y.P. Huang, W.S. Chen and W. Lee. 2007. Electrical and electro-optical properties of nematic-liquid-crystal–montmorillonite-clay nanocomposites. *Jpn. J. Appl. Phys.* 46: 7368.
[306] Huang, Y.P., T.Y. Tsai, W. Lee, W.K. Chin, Y.M. Chang and H.Y. Chen. 2005. Photorefractive effect in nematic—clay nanocomposites. *Opt. Express* 13: 2058.
[307] Tsai, T.Y., Y.P. Huang, H.Y. Chen, W. Lee, Y.M. Chang and W.K. Chin. 2005. Electro-optical properties of a twisted nematic-montmorillonite-clay nanocomposite. *Nanotechnology* 16: 1053.

[308] Jeong, E.H., K.R. Sun, M.C. Kang, H.M. Jeong and B.K. Kim. 2010. Memory effect of polymer dispersed liquid crystal by hybridization with nanoclay. *Express Polym. Lett.* 4: 39.

[309] Huang, Y.P., Y.M. Chang, T.Y. Tsai and W. Lee. 2009. H-PDLC/Clay nanocomposites. *Mol. CrystLiqCryst.* 512: 2013.

[310] Chang, Y.M., T.Y. Tsai, Y.P. Huang, W.S. Cheng and W. Lee. 2009. Polymer-encapsulated liquid crystals comprising montmorillonite clay. *J. Opt. A-Pure Appl. Opt.* 11: 024018.

CHAPTER 3

Potential Applications of Nanoparticles Aided Liquid Crystals

1. Introduction

For several materials, the transition from liquid to solid phase is a series of mesophases known as liquid crystals, rather than a single step (LCs). LCs are self-organized anisotropic fluids that exist thermodynamically between the isotropic liquid and the crystalline phase, showing both the fluidity of liquids and the long-range lattice organisation observed exclusively in crystalline solids [1–3]. LCs are made up of anisotropic construction blocks (typically in the form of rods or discs) that spontaneously arrange themselves in a specified direction called the director n [4]. The director of a nematic LC, the simplest phase of LC with only orientationally ordered molecules, is usually spatially changed continuously but randomly over large spatial extensions in the absence of an external alignment force (except for defects, where the director may vary suddenly and dramatically) [5, 6]. LCs are often divided into two categories: thermotropic LCs and lyotropic LCs [7–10]. The morphology of the constituent molecules in thermotropic LCs is commonly classified as calamitic for rod-like, discotic for disk-like, and sanidic for brickor lath-like molecules [11]. Because LC may self-assemble and so preserve topological defect structures due to their orientational elasticity, they have the potential to be used in modern display technologies [12]. Recent advances in the theoretical knowledge of electrical characteristics have aided in the development of advanced LC materials [13]. It has also been established that by altering the polymerization conditions, different morphological characteristics of LCs can be obtained [14]. It has been calculated that around one out of every 200 organic substances exhibit mesomorphic characteristics. X-ray diffraction, refractive index, thermodynamics, dielectric, spectroscopic, surface tension, and proton resonance are a few ideas and models of molecular organisation that have undergone extensive research. These birefringent mesogenic compounds are utilised to detect various analytes and biomolecules as well as their binding interactions with other molecules because they alter the director's (direction of preferred orientation) orientation and may be seen optical [15]. While researching the derivatives of cholesterol, an Austrian botanist named Friedrich Richard Reinitzer made the initial discovery of the LC phase [16]. He saw that the clear white cholesterol benzoate dissolved from powder to a murky liquid around 145.5 degrees Celsius (bright blue-violet color). And the turbidity disappeared and it became a transparent liquid when it was heated further up to 178.5 C (pale blue color). The LC molecules lack short positional ordering but exhibit long-range orientational ordering (a crystalline-like feature) (liquid-like property). Temperature-dependent melting during the transition causes this mesophase's hue to vanish, though it can occasionally persist in the crystalline phase due to rapid cooling. Otto Lehmann, a German physicist, hypothesised that the LC substance had some optical qualities of solids and flowed like isotropic liquids after studying it under a polarising optical microscope (POM) [17]. He first gave it the term "flowing liquid," and then later changed it to "liquid crystal" [18].

There is a varied area of applications of LCs that have significant promise for basic science and technology such as physics, material science, chemistry, space, engineering, biology, food science, and pharmacology, etc. [19]. Nematics and smectics are commonly employed in LCDs such as television displays, digital watches, dynamic information billboards mobile phones, laptops, tablets, digital projectors, clocks, pocket calculators, information displays, computer monitors, virtual reality, etc. [20]. A homogeneous, flat electronic display screen is employed as a visual display by the LCD visual display equipment. It produces a visual impact on the screen with the use of LCs [21]. In order to generate images with the help of LCs, LCD must be exposed to sunshine or ambient light. The LCD is composed of either an active matrix display grid or a passive matrix display grid. The passive matrix possesses a configuration of a set of conductors and the pixels are situated at the intersection over the grid. For the pixel to have the right luminance, the forwarded current travels via two conductors. Less current is needed when using an active matrix, also referred to as a thin-film transistor (TFT) display, as a transistor is positioned at the junction of each pixel. It has one data line and a parallel column current line, which improves the speed at which the screen refreshes [22]. Due to its extremely low power consumption, LC is mostly used in display technology. Beyond the world of displays, LCs can also be employed in non-display devices like memory, holography, optical ones like phase-shifting interferometers, LC lasers, tunable filters, optical computing, etc. because of their wide range of distinctive and alluring qualities [23]. Additionally, LCs can be utilised as a temperature sensor [24]. A thermometer using LC that changes colour with temperature is known as an LC thermometer, temperature strip, or plastic strip thermometer [25]. Mixtures of LCs are enclosed in separate partitions and the number of partitions show temperatures dependent on the quantity of heat present. Batteries can employ LC as a charge indicator [26]. The theory behind liquid crystalline thermometers is the same. As the electric current moves through the medium, heat is produced. Thus, the change of the respective orientation of the layers creates a colour change. It can be used to find the circuit's weak connections as well.

Numerous other applications for LCs exist, such as mechanized material testing. Radiofrequency waves in waveguides can be imagined using it [27]. Another application of LC is optical imaging. Different disorders are diagnosed and treated using this method. The computer coupled through a camera can display tumours and fractures as rays pass through damaged areas using this technology [28]. Images are displayed on the screen and ultraviolet and infrared rays are employed to diagnose illnesses. Helmets and bulletproof vests are made from long-chained stiff polymers with strong intermolecular interactions that are oriented parallel to one another. In addition to these uses, there are some impressive novel LC applications in metamaterials, photonics, functionalized polymer fibres, sensing, and diagnostics [29]. The probable interaction between functionalized nanoparticles (NPs) and LC to enhance tunable physical properties that enable development of effective biosensing necessary for health wellbeing is highlighted in this chapter keeping the aforementioned factors in mind. In the field of biosciences and medical research, the NPs-dispersed LCs have demonstrated ground-breaking developments to obtain optical and electrical signaling. In the context of biological research, NPs enabled LCs to display special optical, electrical, and molecular characteristics that are missing in pure LCs. The fabrication of such modified self-assembled NP-LCs nanosystems is simple, and their performance can be easily tuned using a host (LCs)-guest (NPs) chemistry-based method. For instance, altering and enhancing certain physical parameters from the original LCs. The LCs are rearranged to create an intriguing material with some additional enhanced qualities.

2. Application of Various NPs Aided LCs in Biosensing Techniques

The use of smart functionalized NPs with desirable electrical and optical properties is advocated in this chapter as a way to improve the sensing performance of LC-based biosensors. In order to obtain effective diagnostics and environmental nonrioting, specialists are paying attention to the contribution of LC in sensor development [30]. To obtain the appropriate phase, which is helpful for creating sensitive sensors, the LC may be easily aligned, and further controlled polarization vial tuning can be stimulated to increase temperature (Popov et al., 2018). In such cases, a strong optical signal amplification will be required for precise spectrophotometer detection. As shown in Figure 3.1(A), 8CB (4-cyano-4'-octylbiphenyl) underwent a regulated phase change over time (0–90 min). This was caused by the creation of a hybrid alignment [31]. By doping a suitable surfactant, such as SDS (Figure 3.1(B)), an LC can be converted into a chiral phase, changing the periodic texture's pitch [32]. A platform like this is highly helpful for detecting the chirality of different biomolecules. Additionally, surface functionality is emerging as a new method for designing and creating sensitive LC-based biosensors for the targeted biomolecule's selective detection. Amphiphilic mesogens can connect with an LC material according to desired functionality, such as a biorecognition molecule. These methods, which are surface chemistry-based, greatly simplify transition alignment, allowing for the selective detection of a desired marker (Figure 3.1(C)) [33]. In addition to surface modification, specific biomolecules can be linked to LC using a cross-linker, such as EDC-NHS and biotin-avidin chemistry (Figure 3.1(D)) [32]. This method is used to create functionality that enables binding with a biorecognition component, such as DNA or an antibody, to detect the targeted biomarker at low and wide levels. The adjustment of an LC structure, alignment, effect of stimulation, and final desired qualities are essential for creating a sensor with therapeutic value. Despite experimental methods, computational science is being introduced much more frequently. This method aids in optimising synthesis pathways, comprehending forecasts, increasing success rates, and evaluating risk considerations. The LC materials that have been optimised are known to display the desired qualities, such as chemo-responsiveness, as seen in Figure 3.1(E) [34]. Apart from computational chemistry, molecular dynamics (MD)-based calculations are used to enhance the physical characteristics of an LC system in a given setting. According to the intended use and environment, MD will simplify the material's structures, such as binding/non-binding sites, while taking all factors into account during simulation. The arrangement of an LC thin film at the interface with a substrate or biomolecules will be revealed by MD in order to create a sensor, as shown in Figure 3.1(F) [35]. The results of LC-based sensing systems are widely used in the development of increasingly sophisticated and effective structures.

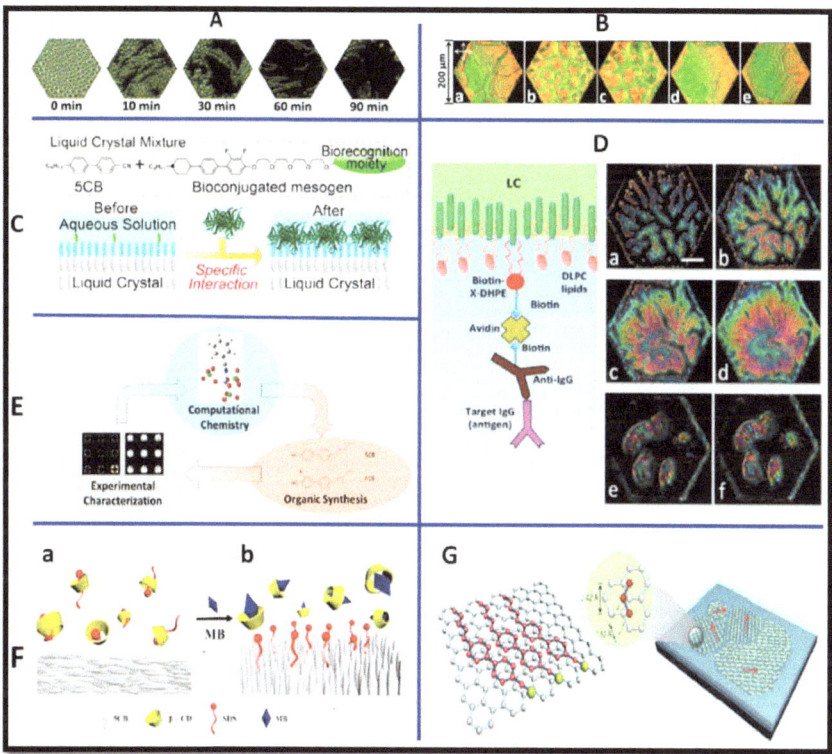

Figure 3.1. Toric focal conic domain, formed by hybrid-aligned 8CB film, coverage as a function of time (0–90 min) at 37°C (A), A 5CB thin film texture coverage with a chiral dopant as a function of water (0–1.89 mM) (B), functionalization of LC for specific binding (C), surface chemistry, i.e., biotin-avidin to achieve specific binding with a targeted biomarker (D) and performance of developed immune sensor using biotin-avidin chemistry. This sensor detected 10 mg/mL of antigen as a function of time ranging from 10 to 50 min (D), Computational chemistry-based chemo-responsive design of a LC (E), Host-guest interaction of b-cyclodextrin (CD) and methylene blue as evaluated using MD (F) and exploring orientation of graphene domain in thin film using a LC (G) [Reproduced from Ref. 32].

With the advent of wearable sensing systems, these new LC sensing materials will be more accessible, sensitive, and adaptable. The need to incorporate nanotechnology into the science of building LC-based smart and efficient sensors is needed for selective diagnoses and sensitive environmental monitoring technology gained popularity by LC-based sensors to maintain targeted future strategy [36].

Biological sensors and drug delivery are improved in numerous ways by the self-assembly of nanoparticles (NPs) in the LC phase. Ferroelectric, metal, semiconductor, and carbon NPs are among the four classes of NPs used to enhance the properties of the host LC material [37–39]. Due to the interaction of the ferroelectric NP's Ps with the elastic forces of the LC, the ferroelectric NP significantly influences the LC by raising the order parameter [40]. One of these fascinating studies used 1D or 2D anisotropic NPs in lyotropic LC to accomplish self-assembly [41]. This soft matter leads to a more sophisticated and enhanced medical technique to identify immune-related bio-actives linked to many forms of diseases that harm humanity, based on identified improved qualities [42]. The improvement of LC inclination disruption in the presence of NPs has been described in many papers, and this allows for the straightforward and amplified detection of a targeted marker utilizing NP-assisted LC based sensors [43]. To easily comprehend the enzymatic activities between cholylglycine hydrolase (CGH) and cholylglycine (CG), Wei et al. recently dispersed nickel (Ni) NPs in LC [44]. As a result, the immobilization of CG on the CGH layer could disrupt the consistent homeotropic alignment created in the LC by Ni NPs. Surface-functionalized nanoscale structures with desired shape, size, morphology, surface charge, and intrinsic optical, electrical, magnetic, and molecular properties will be made possible by nanotechnology [45–47].

These nano-structures/systems can easily combine and align with an LC due to their small size, creating a new phase or class of materials, such as hybrid composites. These innovative qualities, such as LC and a particular nanostructure, will not be present in progenitors of this new type of sensing material. The biological entity is able to connect to the surface of the LCs with ease because NPs improve the local orientational order parameters of the LCs [48]. This idea would also aid in enhancing the selectivity, compatibility, and efficiency of various contemporary disorders like cancer [49]. Incorporating additional techniques, such as the Photonic Immobilization Technique for antibody functionalization on electrodes, can let LC identify more analytes in the future. This method will be useful for detecting light molecules and offer adaptability, portability, and rapid response [50]. The optical, electrical, and magnetic properties that are beneficial to construct a sensor with high sensitivity, wide detection range, low detection range, and high stability may be easily regulated in the hybrid system of an LC and nanostructure. In this regard, research has been done to examine the relationship between LC and nanostructures such as tungsten diselenide, molybdenum diselenide, and carbon nanostructures. For instance, by imaging the birefringence of a nematic LC caused by a strong-stacking contact, LC has been investigated to comprehend the alignment of the 2D graphene sheet (Figure 3.1(G)) [51]. Surface effects have a significant influence on how LC molecules align. Therefore, when the LC cell is seen under POM, a slight shift in the surface causes the LC molecules to be oriented in a different orientation. This results in some transmitted light intensity through the LC cell. On a sheet of ordinary glass, the substrate is initially chemically functionalized. No light is passed through the sample cell during this process. The targeted diseases' enzymes are then added to the glass slide. When viewed under POM, a minor light transmission is discernible through the sample cell. Additionally, once the surface has been merged with NPs, the light can be seen at its brightest when it incidentally strikes the LC materials [52]. First, Liu and colleagues present a straightforward method for the highly sensitive detection of DNA utilising gold nanoparticles (AuNPs) [53]. The surface density of DNA strands changes significantly when AuNPs interact with them. Additionally, thin LC layers are amplified when they are interrupted, which causes the LC molecules to transition from being homeotropic to tilted and causes optical textures to be seen rather than a dark background. It offers great selectivity and a much lower 0.1 pM detection limit. Additionally, the variation of the birefringent domain with respect to the concentration of the targeted DNA [54]. A LC-based biosensor has been described to detect L-tyrosine (Tyr), an amino acid needed by cells to build proteins [55]. Tyr activity directly affects the melanin's capacity to absorb light and protect skin cells from UVB radiation, which lowers the risk of cancer and is used to treat many congenital conditions including albinism. Tyr detection is therefore crucial [56]. The birefringent texture in the optical images was enhanced as the Tyr concentration rose, demonstrating the massive production of l-DOPA (L-3,4-dihydroxyphenylalanine) to form large-size AuNPs that alter the surface topology of LC cells and cause the LC molecules to transition from homeotropic to tiled. As a result, the birefringent property of LC under crossed polarizers is used to develop and create an LC biosensor based on the catalytic enhancement of AuNPs for the detection of Tyr. Particularly at low Tyr concentrations, this biosensor functions adequately. Acetylcholine (ACh), an organic compound that regulates the functioning (in the brain and body) of humans and animals as a neurotransmitter and acetylcholinesterase (AChE) inhibitor, has been detected using an LC biosensor based on the enzymatic development of AuNPs [57]. The optical pictures of the birefringent texture diminish as the ACh concentration rises from 0.015 to 1.5 mmol/L when ACh is introduced in various quantities to the AuNPs growth solution [52].

A brand-new LC-based sensor for thrombin detection was created using nickel nanospheres (NiNS). By doping NiNSs, the homeotropic alignment of 5CB was quickly attained. In order to increase the disruption of the LC orientation and the resulting orientational transition in LCs, a sandwiched system of aptamer-functionalized AuNPs was created. It is simple to see this transition, and thrombin concentrations between 0.1 and 100 nM were found [30].

The primary inspiration for creating biosensors came from the need to manage better healthcare. It can be used as an immediate corrective measure to give a patient's metabolic status a more accurate estimate. It is manageable due to its modest size [58–60]. Since LC-based biosensors are inexpensive, they can also be utilized for the detection of various diseases and metabolites, including blood pressure, body temperature, heart rate, cancer, stroke, and multiple sclerosis. The quantitative estimation of various biologically significant compounds in body fluids, including glucose, cholesterol, urea, endotoxin, DNA molecules, cadmium, bile acids, tuberculosis (TB), ATP and others, is successfully carried out using biosensors. LC-based biosensors have uses outside of clinical diagnosis and biomedicine, including in the pharmaceutical industry (estimation of drug residues in food, such as antibiotics and growth promoters, drug recognition, and assessment of biological activity of newly synthesized compounds), fermentation control, food and beverage analysis, industrial effluent control, pollution control, mining, industrial, and toxic gas examination, as well as military applications. Researchers can now more thoroughly understand the structure and interface phenomena thanks to a variety of LC biosensor approaches that have been disclosed (Rogers, 2000). Therefore, biosensor technology has a huge potential and revolutionizes the study and management of biological systems.

3. Applications of Gold NPs Doped LC Composites in LCD Systems

LC, monomer, and initiators are components of the formulations for traditional PDLCs. The application of additives to the basic components were also investigated by several researchers in order to give higher device attributes and overcome perceived limitations like high operating voltages or poor response times. The anchoring energies at the polymer/LC interface, the medium's dielectric constant, or the refractive indices of the LC or polymer matrix are the main effects of NP addition. Operating voltage, transmission characteristics, and response times are all affected as a result. Different NP kinds are now being employed to create PDLC films with enhanced properties [61]. An improved EO performance of the PDLC films is ascribed to the incorporation of metallic NPs. Because surface plasmon excitations at metal-LC interfaces result in an improvement in local electric fields, Au and Ag-doped films in particular exhibit high CR and low driving voltages [62, 63]. The GNPs capped LCs have drawn a lot of interest in display technology during the past 20 years, garnering a lot of attention from numerous research organizations. Because of this, a significant amount of research has been done in this particular domain of study. In this part, discuss and provide summaries of some of the most popular methods for creating GNPs-capped LC displays. Table 3.1 lists the different GNPs and LCs that were employed in the different LC display composites. There is a brief discussion of the several

Table 3.1. The detailed characteristics of LCs and polymer materials capped with GNPs applied in the different LC related display composites.

S. No.	Gold nanoparticles (GNPs)		LC Host Materials		EO response/results	References
	GNP Diameter (nm)	GNP Doping	LC Name	LC Type		
1	3 nm	1:1 mol ratio	10-[(trans-(4-pentyl cyclohexyl) phenoxy)] decane-1-thiol (5CP10SH)	Monotropic: a nematic phase and a smectic phase on cooling	Fluidity at the mesophase and optically isotropic state	[64]
2	1.3–10 nm	5 wt%	LC1-LC3	Nematic	Chirality was transferred to the nonchiral LC. Uniform stripe textures/patterns separated by areas of homeotropic alignment formed.	[65]
3	1.4–3.5	5 wt%	4-pentyl-4cyanobiphenyl (5CB)	Nematic	First experimental proof of the usefulness of GNP's as chiral dopants for N-LCs. The non-chiral Au cluster, at a macroscopic level, no relationship between the cholesteric fingerlike textures and chirality was found.	[66]
4	10, 25	0.1 wt%	CHS1: based on three phenyl pyrimidenes homologs and a chiral component	Ferroelectric	A significant decrease of the switching time and the rotational viscosity, independently on the Gold Nanorods (GNRs) diameter. Conversely, the relaxation frequency and the dielectric strength of the Goldstone mode was strongly dependent on the GNRs diameter.	[67]

5	7	0.03–0.54 volume fractions	4-n-octyl 4_cyanobiphenyl (8CB)	Smectic	Due to the one-dimensional quasi-long-range solid-like structure's restriction on nanoparticle motion across layers and their elasticity-mediated interactions, which prevent irreversible aggregation and increase the stability of the resulting nanoscale dispersions in thermotropic smectic liquid crystals.	[68]
6	6	0.2 wt%	4-cyano-4-npentylbiphenyl (5CB)	Nematic	The threshold voltage for an electric field-induced homeotropic director alignment was reduced by about 25% while the nematic-to-isotropic phase transition temperature increased by just 0.4°C.	[69]
7	10	1 wt%	Nematic Liquid Crystal (NLC) (E-7)	Nematic	GNPs lower switch-on voltage and also lower the relaxation frequency with applied voltage.	[70]

methods and processes that have been employed in the past and present to create the GNPs-capped LC display films.

3.1 GNPs Capped in Nematic LCs

Double-melting performance was seen in one of the earliest experiments for GNP-capped liquid-crystalline composites [64], and mesomorphic response of the doped GNPs indicated the interaction between mesogens on the surface of various GNPs. It was hypothesised that GNPs' mesomorphic behaviour may be used to alter the aggregates' morphology (lattice structure, interparticle separate, etc.). A uniform stripe creation was described in two different ways in one of the pioneering works of H. Qi and T. Hegmann [65], where they described the formation of periodic stripe textures in NLCs that have been doped with functionalized GNPs (chiral, and achiral alkylthiolates) in small quantities (5% wt percent): (i) the chiral alkyl thiol-functionalized GNPs transfer their chirality to the chiral host NLC, and (ii) the creation of topological flaws through chiral/achiral GNPs. It was also discussed how LC doped GNPs affected the performance of electrical conductivity and dielectric constant [71]. The mixture was made up of a commercially available, room-temperature nematic substance that had been doped with alkylthio functionalized GNPs (15–20 nm, 5 wt%). The nematic-isotropic transition temperature (TN-Iso) was discovered to depend on the quantity of GNPs based on the observed results. The addition of GNPs enhanced the composite's electrical conductivity, but the researchers also discovered a slight loss in anisotropy. Using chemically functionalized GNPs doped into LCs, localized surface plasmon resonance (LSPR), which can be tweaked to change the ordering of the LCs close the GNPs (~ 10 to 40 nm), proved useful for detecting the changes [72]. Additionally, the LSPR behaviour of the NPs and the LC's far-field orientation were contrasted. It gave information about the nanoscopic causes of the LCs' bulk anchoring characteristics. The research showed that it is possible to investigate how the functions of NPs may influence the local ordering of LCs using the LSPR features of NPs. The design guidelines for structure-property association for the NPs capped with an organic monolayer were investigated by Mehl and colleagues [73, 74]. When mesomorphic features were determined by OPM and DSC, the scientists discovered that the GNPs bearing side-on functionalized mesogenic units have a peculiar nematic marble and schlieren texture. The efficiency of gold nanoclusters as chiral dopants for NLCs was the topic of an experimental study by Qi et al. [66] that validated the chiral shift from chiral GNPs to the achiral NLCs via induced circular dichroism (ICD). It is interesting to note that there is no association between chirality and stripe texture, especially for the achiral GNPs cluster [66]. The Freedericksz Transformation and an unusual dual alignment mode have also been seen by the authors in alkylthiolfunctionalized GNPs-doped NLCs [75]. Lower threshold voltages and higher dielectric constants were found in the work for the pure NLCs in both thermal and field-dependent modes. According to the results, functionalized gold nanoclusters could be used to create improved or new LC display applications, such as optical biosensor designs. Additionally, by doping GNPs into the composites, several methods were used to modify the alignment of NLCs [76, 77]. It was determined that the concentration, composition ratio, size, and carbon chain length of the alkyl thiolate capping's, the order parameter, ordering around (or in the bulk) the NPs present at the interface, as well as the sample fabrication environment and sample preparation methods, all have an impact on the alignments of NLCs. In their subsequent effort, which was published in 2009 [76], the group described an investigation into the alignment and miscibility effects of single layer GNPs doped with cyanobiphenyl LC in nematic cyanobiphenyl (CB)-related LC hosts and contrasted their results with those of NLCs doped with alkyl thiol-capped GNPs. In contrast to those based on alkyl thiol-capped GNPs, the authors found that the CB related LC-capped GNPs were less miscible in NLCs [76]. Through a straightforward two-step synthesis process, Khatua and colleagues [69] have successfully replaced decanethiol with 4-sulfanylphenyl-4-[4-(octyloxy)phenyl] benzoate (SOPB) capped GNPs (size of 6 nm, and doping 0.2 wt percent) nematic phase solubility of 4-Cyano-40-pentylbiphenyl (5CB) The threshold voltage (Vth) for the Freedericksz transition fell by about 25%, while the temperature of the TN-Iso phase transition increased somewhat. It states that the SOPB capping agent was used to increase the solubility at a higher concentration of GNPs. The results were undoubtedly very helpful for the dissolving of nanoparticles with adjustable plasmonic shape and size in LC composites. NLC composites with increased anisotropic electrical conductivity, dielectric properties, and elastic constants were studied as a result of the addition of gold nanorods (GNRs) [78]. The observations of the elastic constant predicted an improvement in the anti-parallel correlation of the molecules in the isotropic phase as well as an increase in the ordering in the nematic phase.

On the basis of the shape anisotropy of the GNRs, the authors further explained their conclusions. Sputter doping of the target material on the LC host was recommended in a study as a straightforward method to create MNP-LC suspensions [79]. Instead of using traditional solid substrates, LC substrates were employed in this study because they can store NPs from the target material sputter doping procedure. It claimed that the technology used was straightforward, quick, and enabled pure production. The research also showed that for the sample made with a twisted hematic cell utilizing 5CB and 5CB deposited with gold NPs, the threshold driving voltage was frequency dependent. There was a noticeable dependence on the frequency predicted by the NP-dispersed 5CB, as well as a decrease in threshold voltage compared to the pure 5CB. To further simplify things, it was reported [80] that alignment of the LCs in LC/NPs dispersions was investigated using fluorescence confocal polarising microscopy (FCPM) and polarising optical microscopy (POM) methods. The results of the 2- and 3-dimensional imaging experiments demonstrated the twist declinations represented by the birefringent stripes caused by the NPs.

Qi et al. thoroughly examined their earlier research on the description of structure- and composition-related linkages connected to LC-NP interfaces [81]. The development of birefringent stripe textures and the introduction of homeotropic alignment of the NLC, which is akin to chiral finger (or fingerprint) textures, were two novel textures for GNP-doped NLCs that the authors observed.

Numerous experimental experiments using GNPs with different core sizes and capabilities were built on this unique texture. According to the research, these novel techniques may open up new avenues for improving the EO properties of LCs used in display technologies. More significant improvements in the electrical conductivity and TN-Iso, as well as a decrease in the dielectric anisotropy and Vth, were noted in NLCs containing functionalized GNPs (2.4 nm in size and 0.1 wt percent doping) [59]. The study's findings indicate that doping on GNPs decreased the effective longitudinal component of the system's dipole moment and, as a result, increased the stability of the nematic phase, resulting in the enhancement in TN-Iso. It has been documented that the addition of GNPs (10 nm) causes significant alterations in the electro-optical (EO) properties of NLCs [70, 82]. The team noticed that the switch-on voltage was reduced with the addition of GNPs. Furthermore, the relaxation frequency decreased when the voltage was applied to the NLC cell. As a result, the relative dielectric permittivity started to decline at the lower frequency of the applied voltage. The changes in the relaxation frequency and Vth in the GNPs-doped NLC composites were also theoretically explained. Milette et al. [83] attempted to create the homogeneous dispersion in LC composites in order to produce GNPs from suitable optical applications with a ligand shell by taking advantage of the notion of the existing contacts between the host LC molecules and GNP organic shells. The derivatization technique created in this study for mesogenic ligands offers a strategy for building bifunctional GNPs through thiol-for-thiol exchange. This method offers good repeatability, a broad compositional range, and a relatively modest amount of valuable functionalized ligand. The results of their experiments supported the value of the 4–5 nm 4-(N,N-dimethylamino) pyridine (DxMAP) system in the production of appropriate GNPs. Later, K. K. Vardanyan and his team [84] produced binary mixtures of a polar NLC with (1–10 wt percent) doping of GNPs to show off their successful achievements. According to them, the dielectric anisotropy doubled with mixes containing 1 to 2 wt percent GNPs, while the birefringence increased by around 50%. At particular GNP concentrations (35%

and 45%), a nematic-IL phase transition point was seen that was C higher than it was at 0 percent concentration [85]. Additionally, the dielectric anisotropy doubled for a concentration of 45 percent and rose by 2.2 times for 35 percent. Reviewing the research on the dispersion of GNPs in several homologues of CB-based LC with various concentrations (nCB with n145, 6, 7) [86] is a continuation of this work. The findings from this work showed that, GNPs' impact on the attributes of different aggregation between GNPs and LC molecules, results into the transformed properties in the materials; particularly provides thermal stability to NLCs. Results of the dispersion of 20 wt% GNPs-doped 5CB at room temperature were provided in further investigations [87]. With the exception of 15 wt%, all GNP concentrations in the studied composites increased their elasticity and dielectric anisotropy. Despite this, the rotational viscosity continued to decline, stopping only at 2%, with no change from the bulk values. As a result, neither the material properties nor the thermal stability of the composite's nematic phase were affected by the doping of GNP. The creation of a thermomechanical nematic LC elastomer (LCE) based actuator with enhanced material response on increasing temperature was reported by the team of Montazami [88], focusing on an important application. At high heating rates, a more than 100% increase in the rate of change of strain was seen when using GNPs-doped LCE actuators. The GNPs-doped LCE actuators' improved response allowed for its use in a variety of applications. The Shuzhen team created a special acetylcholinesterase (AChE) LC biosensor using the enzymatic development of GNPs for the detection of AChE inhibitor and acetylcholine (ACh) [52]. In a work published in 2012 [89], the evolution of self-assembly of a series of GNPs capped with mesogenic units (ligands) and various ratios of alkylthiols as co-ligands was demonstrated. It states that the presence of mesogens caused GNPs to pack distantly in an orderly manner. The study focused on the relationship between the size of the co-ligands, the number of mesogens connected to the GNPs, and the GNPs' responses to 2- and 3-dimensional building. It was discovered that an improvement in optical nonlinearity occurred when LCs were doped with modest amounts of GNPs coated with tetra octyl ammonium bromide (TOAB), leading to an intriguing improvement in the two-beam coupling effect [90]. The composites of GNPs and a weakly polar NLC exhibiting low-frequency director relaxation were the subject of the first thorough calorimetric, frequency-dependent anisotropic permittivity and conductivity measurements, which were reported in 2014 [91]. Their findings show that the nanoparticles greatly reduced the TN-Iso and the associated transition entropy. Additionally, the composites' conductivity doubled, and a percolation scaling law—typically found in metal-based polymer nanocomposites—showed how conductivity depended on concentration. By employing the spectrophotometric approach to examine the response of the absorption peaks of the surface plasmon polariton (SPP), a similar analytical research [92] was done on the GNPs doped LC director. By using linearly polarised light that was directed at 0, 45, and 90 degrees, which corresponds to the direction of rubbing, the incidence produced contact with the LC composite. The anisotropic variation at the height of the SPP absorption peak was seen here thanks to light transmission. Evidently, the GNP doped LC molecules' anisotropic motions were seen. Silane conjugation was used to create chemically and thermally resistant LC silane-functionalized GNPs [93]. In their research, the electro-optic response, optical response, and thermal response of composites based on NLC at different GNP concentrations were examined. In their work, Urbanski and team [94] described and discussed how the interaction of nanoparticles with the aligned layers of the composite electro-optical samples significantly complicates the electro-optical properties of LC/NP composite systems.

The elastic constants, dielectric permittivities, pretilt angles, and polar anchoring energy were among the different characteristics that the authors used in their work to determine and compare with experimental results. With a minor reduction in the dielectric anisotropy, the simulated results confirmed the experimental findings in terms of elastic constants. According to one of the related investigations, nematic LC/NPs significantly enhance display properties and increase the temperature range of the nematic phase when used at room temperature [95]. According to the group, the doped GNPs effectively alter the local ordering of LCs, which has an impact on the bulk phase's display and material properties. The usage of azobenzene thiol doped GNPs was claimed to have enabled the reversible orientation control of LCs by light irradiation approach later in 2015 [96]. The azobenzene-based thiol, designated AzoSH, has two functions: (a) this made it possible for GNPs to be mixed uniformly in NLC and (b) this served as a photo-isomerizable substance to customize how well the functionalized GNPs produced LC molecule alignments. The LC nanocomposites underwent reversible alignment change under UV/Vis irradiation as a result of the azo groups being trans-cis photo-isomerized on the surface of GNPs, making the produced device ideal for display applications. Determining the range of chirality from a nanoparticle surface to the bulk was the subject of an intriguing study by Mori et al. [97]. Three enantiomeric pairs of chiral ligand-capped GNPs were created by the team. Here, the chirality transfer mechanism was triggered and customized using GNPs of different curvature, size, and ligand density. In order to determine the effectiveness of the binaphthyl ligand doped GNPs' chirality shift, the helical pitch values and the helical twisting power (HTP) of the induced chiral nematic phase and chiral GNPs were calculated. Surprisingly, they discovered that while chirality shift efficiency deteriorated with decreasing particle size, the HTP strengthened with increasing particle diameter.

The nonvolatile memory effect of GNPs-doped orthoconic smectic LC mixtures was studied using various sizes and types of polymer capping, and memory effects were observed in all prepared samples with various types of NPs and LC, highlighting the significance of the host LC, chemical nature, and GNP characteristics. An ideal GNP concentration for the electro-optical properties of GNP-doped nematic LC cells (NLCs) was around 0.07 wt% [98]. Response time and Vth of the cell decreased by 28% and 21%, respectively, when compared to the pristine LC cell at such concentration. Because of the LCs' reduced rotational viscosity, the authors found that adding modest amounts of GNPs reduced the LC cell's response time. Additionally, the doped GNPs reduce the LC device's reorientation angle during voltage switching, which cuts down on response time. The doping of significant amounts of GNPs slowed the LC cell's reaction time because the extra GNPs were networked and accumulated within the cell. As a result, it interfered with the LC alignment and made the LC realignment difficult. A study on light propagation in photonic crystal fibres combined via 2-nm GNPs doped with a 6CHBT; NLC (varying from 0.01 to 0.5 wt%) was recently conducted in 2019 [99]. Their results demonstrated a significant improvement (about 50%) over the prior findings (from 2017) for the rise and fall times for photonic LC film infiltrated with the same LC and capped with 0.5 percent GNPs, with an 82% reduction for the rise time. They also noticed a reduction in the Freedericksz threshold voltage of up to 60% for the samples they used. A study on liquid crystalline composites made with GNPs and organic ligands employing plasmonic resonance was also presented [100], and the authors found that temperature changes and light insertion may control the optical properties of the composites.

3.2 GNPs Capped in Cholesteric LCs

GNPs were used to modify the properties of cholesteric liquid crystals, where it was shown that capping GNPs increased the temperature of the blue phase by 0.5–5.0 C in order to stabilise the blue phase [101]. The outcomes show that the GNPs' aggregation in lattice disclinations stabilized the entire blue phase composite. The tunable refractive index-independent polarization in the GNP-stabilized blue phase composite was investigated in a different study [102]. The research showed that the nanoparticle-doped blue phase II inhibited the field-induced BP-cholesteric shift. Greater applied voltages resulted in a wider refractive index (RI) tuning range because the LC remained optically isotropic. BP system with extended blue phase range and electrically thermally adjustable optical properties was the subject of a study on gold nanorods (GNRs) [103]. The GNR scattered BPLC composite, according to it, displayed a low switching voltage but with a little increase in response time.

Later in 2013, Yoshida's team [104] investigated how the phase of liquid crystals' cholesteric phase and blue phase affected the dispersibility of spherical GNPs. The researchers concluded that the structural differences between each LC phase were the cause of the discrepancy in dispersibility. According to Liu and colleagues [105], cholesteric LC indicates the characteristics of photonic type crystals as well as a photonic bandgap. In that study, laser dye and GNPs were combined, and the authors reported cholesteric LCs with a defect mode, which resulted into useful feature of localized surface plasm on impact produced through the GNPs. The luminescence basically increased after that [105].

3.3 GNPs Capped in Smectic LCs

The doping of citrate ion-capped GNPs has been thoroughly studied using the chiral smectic C phases among all the other smectic LC phases, in accordance with the scientific relevance of the SmC phase. Kaur and colleagues [106] demonstrated the improved electro-optical and textural features employing the GNPs through the composites of ferroelectric liquid crystal in noteworthy work in this category (FLC). The outcome shows an optical tilt as a five-times rising and lowering voltage threshold of 0.1 V. Additionally, the composite of doped FLC showed the effect on memory. Results were shown in two different methods. The first was an improvement in the FLC composite as the intrinsic electric field was seen, which was probably caused by the insertion of GNPs into the samples under study. In the second method, electron wave oscillations interacted with electromagnetic wave oscillations before being transmitted by FLC molecule from the incident photon light. The scientists looked at the outcomes of a work on citrate ion-capped GNPs that focused on distorted helix ferroelectric liquid crystals (DHFLCs) [107]. The stabilisation of DHFLC helix deformation and the charge transfer process brought on by the electric field were credited by the researchers as the causes of their newly discovered memory effect.

A study by Kumar et al. [108] revealed a nine-fold increase in the photoluminescence (PL) intensity in a newly constructed DHFLC composite by GNPs doping, which might be seen as useful in the realization of PL-based LCDs in addition to the better EO response that was previously discussed. According to the authors, the creation of strong electromagnetic fields near to the GNP surfaces improved the photoluminescence (PL) of the doped LAHS19 composite GNPs. The creation of a tunable optical bulk metamaterial used Au-nanospheres dispersed in a smectic LC, according to a study of a similar nature [109]. The impact of the GNRs' size on the EO and dielectric response of ferroelectric LCs was examined by Podgornov and colleagues [110]. The amount of unrestrained polarization, operational duration, and rotational thickness all significantly decreased, according to the researchers. However, regardless of the GNRs diameter, there was no change in the tilt direction of the GNR-FLC composites. They explained this behavior by pointing to the FLC's enhanced internal electric field, which was made possible by the presence of GNRs [110, 111]. It was performed to improve the coating structure and colloidal durability of the smectic LC nanoparticle scatterings [68]. The examination of actual data and computer models shows that SPR spectra show excellent interparticle partitioning. Even at higher concentrated suspensions, the spectra demonstrate the significant interaction between distant metal NPs in the aggregate of smectic lamellae. Due to layer distortions in the LC-based bulk inclusions, additional NPs in thin films didn't aggregate but instead changed the free surface profile of the film. During the SmC phase of KCFLC 7S LC capped through GNPs, Joshi et al. [37] determined the characterization and dielectric relaxation based on a low-frequency mode method. The researchers discovered that KCFLC 7S is capable of homeotropic alignment on glass surfaces coated with ITO. The highlights of the freshly discovered relaxation mode were demonstrated by the bias-dependent dielectric and temperature characterizations that depend on undoped and GNPs-doped FLCs.

A ground-breaking work examined the effects of the GNPs' chain length and bound LC functional groups on the EO and dielectric properties of alkyl thiol-doped GNP-FLC composite cells [111, 112]. With increasing GNP concentration, it was discovered that EO response characteristics including unconstrained polarization, rotational viscosity, and operating time of the composites all decreased.

3.4 Applications of GNPs in PDLC Systems

A proof for reasonably large-scale evidence for reasonably large variations in the EO characteristics of a PDLC as a result of the insertion of small amounts of 14-nanometer-diameter GNPs [113] was presented. The chemical reactions between $HAuCl_4$ and $Na_3C_6H_5O_7$ were carried out, and sodium citrate reduced gold ions to NPs, which were then synthesised and dispersed in water by GNPs. Based on the concentrations of NPs and the applied electric field, the doping of GNPs reduced the Vth by 50% and increased optical transmission in their work. The frequency response of the film was likewise affected by the doping of GNP into PDLC. Their observed results showed that electric fields in GNP-based PDLCs at metal/dielectric interfaces in the composite material were mostly produced by surface plasmon excitations. Investigated and compared were the characteristics of H-PDLC gratings capped with Ag and Au nanoparticles [114]. The reported work shows that the grating's diffraction efficiency has increased, and the Vth and response time have decreased due to the doping of silver and gold NPs. Un-doped, Ag-doped, and Au-doped grating are the first three. As a result, in this investigation, gratings with silver NP doping had a substantially stronger reaction than those with gold NPs. This is likely because the SPR wavelength of silver NPs in PDLC is much closer to the exposure laser than that of gold NP.

In addition, the AFM images demonstrated enhanced Ag-doped grating phase separation and revealed that the Ag-doped samples' grating configuration was softer than that of the gold NP-doped samples and undoped samples [114]. Using the PIPS technique, such GNPs were doped into a polymer and LC blend [63]. The effect of GNPs on the performance of the PDLC composite's dielectric relaxation in the 20 Hz to 10 MHz range was examined by the authors. According to the observed results, the GNP coated PDLC films showed two distinct relaxation peaks, but the plain PDLC film only displayed a single relaxation process. The relaxation peak of low frequency was found in the GNP-PDLC film at about 24 kHz, whereas the peak of high frequency was discovered in the 1 MHz area [63]. Finally, it was discovered that the UV-Vis spectra of both types of composites remained unaffected. The tetrahertz (THz) birefringence anisotropy of the GNPs-doped PDLC was experimentally studied in one of the most recent research [115]. The scientists used the terahertz time domain polarisation spectroscopy (THz-TDPS) technology to further explore the surprising effect on the high AC electric field [116]. Three different types of samples— pure PDLC, GNP-doped PDLC, and the precursor without UV— were made for the purpose [115]. This constancy of PDLC doped with GNPs behaviour to the applied electric field was appropriate for applications in adjustable THz stage shifters. Second, the relaxation frequency of PDLC was around 530 kHz at a GNP concentration of 0.2 wt%, which was more than twice as high as that of pure PDLC and four times as high as that of the precursor combination (without UV). The faster relaxation responsiveness of LC molecules to the applied electric field was demonstrated at the higher relaxation frequency.

References

[1] Tschierske, C. 2012. *Liquid Crystals: Materials Design and Self-Assembly.* Springer Science & Business Media: Berlin, Germany, Volume 318.

[2] Bisoyi, H.K. and S. Kumar. 2011. Liquid-crystal nanoscience: An emerging avenue of soft self-assembly. *Chem. Soc. Rev.* 40: 306–319.

[3] Kato, T., N. Mizoshita and K. Kishimoto. 2005. Functional liquid-crystalline assemblies: Self-organized soft materials. *Angew. Chem. Int. Ed.* 45: 38–68.

[4] Ohzono, T. and J.I. Fukuda. 2012. Zigzag line defects and manipulation of colloids in a nematic liquid crystal in microwrinkle grooves. *Nat. Commun.* 3: 701.

[5] Chuang, I., R. Durrer, N. Turok and B. Yurke. 1991. Cosmology in the laboratory: Defect dynamics in liquid crystals. *Science* 251: 1336–1342.

[6] Pieranski, P., B. Yang, L.-J. Burtz, A. Camu and F. Simonetti. 2013. Generation of umbilics by magnets and flows. *Liq. Cryst.* 40: 1593–1608.

[7] Dierking, I. and S. Al-Zangana. 2017. Lyotropic liquid crystal phases from anisotropic nanomaterials. *Nanomaterials* 7: 305.

[8] Saliba, S., C. Mingotaud, M.L. Kahn and J.-D. Marty. 2013. Liquid crystalline thermotropic and lyotropic nanohybrids. *Nanoscale* 5: 6641–6661.

[9] Laschat, S., A. Baro, N. Steinke, F. Giesselmann, C. Hägele, G. Scalia, R. Judele, E. Kapatsina, S. Sauer, A. Schreivogel et al. 2007. Discotic liquid crystals: From tailor-made synthesis to plastic electronics. *Angew. Chem. Int. Ed.* 46: 4832–4887.

[10] Hegmann, T., H. Qi and V.M. Marx. 2007. Nanoparticles in liquid crystals: Synthesis, self-assembly, defect formation and potential applications. *J. Inorg. Organomet. Polym. Mater.* 17: 483–508.

[11] Dierking, I. 2003. *Textures of Liquid Crystals.* John Wiley & Sons: Hoboken, NJ, USA.

[12] Salamon, P., N. Eber, Y. Sasaki, H. Orihara, A. Buka and F. Araoka. 2018. Tunable optical vortices generated by self-assembled defect structures in nematics. *Physical Review Applied* 10: 044008.

[13] Naemura, S. 1999. Electrical properties of liquid-crystal materials for display applications. *Mater. Res. Soc. Symp. Proc.* 559.

[14] Filpo, G.D., R. Cassano, L. Tortora, F.P. Nicoletta and G. Chidichimo. 2008. UV tuning of the electro-optical and morphology properties in polymer-dispersed liquid crystals. *Liq. Cryst.* 35: 45–48.

[15] Lee, M.J., C.H. Lin and W. Le. 2015. *Proc. of SPIE* 9565: 956510–956512.

[16] Hamley, I.W. 2007. *Introduction to Soft Matter: Synthetic and Biological Self-Assembling Materials.* John Wiley & Sons, England.

[17] Kumar, S. 2011. *Chemistry of Discotic Liquid Crystals: from Monomers to Polymers*, first ed. CRC Press, New York.

[18] Singh, S. and D.A. Dunmur. 2002. *Liquid Crystals Fundamentals.* World Scientific Publishing, Singapore.

[19] Woltman, S.J., G.D. Jay and G.P. Crawford. 2007. Liquid-crystal materials find a new order in biomedical applications. *Nat. Mater.* 6: 929.

[20] Vass, D.G., W.J. Hossack, S. Nath, A. O'hara, I.D. Rankin, M.W.G. Snook, I. Underwood, M.R. Worboys, M.S. Griffith, S. Radcliffe, D. Macintosh, J. Harkness, B. Mitchel, G. Rickard, J. Harris and E. Judd. 1998. A high resolution, full colour, head mounted ferroelectric liquid crystal-over-silicon display. *Ferroelectrics* 213: 2019–2218.

[21] Brown, G.H. 1973. Properties and applications of liquid crystals. *J. Electron. Mater.* 2: 403–430.

[22] Troccoli, M.N. and M.K. Hatalis. 2013. Active Matrix Display and Method. U.S. Patent No 8,390,536B2. U.S.

[23] Sohn, J.I., W.K. Hong, S.S. Choi, H.J. Coles, M.E. Welland, S.N. Cha and J.M. Kim. 2014. A novel high-sensitivity, low-power, liquid crystal temperature sensor. *Materials* 7: 2044–2061.

[24] Algorri, J.F., V. Urruchi, N. Bennis and J.M. S´anchez-Pena. 2014. A novel high-sensitivity, low-power, liquid crystal temperature sensor. *Sensors* 14: 6571.

[25] Moreira, M.F., I.C.S. Carvalho, W. Cao, C. Bailey, B. Taheri and P. Palffy-Muhoray. 2004. Cholesteric liquid-crystal laser as an optic fiber-based temperature sensor. *Appl. Phys. Lett.* 85: 2691.

[26] Schmidt-Mende, L., A. Fechtenkott, K. Mullen, E. Moons, R.H. Friend and J.D. MacKenzie. 2001. Self-organized discotic liquid crystals for high-efficiency organic photovoltaics. *Science* 293: 1119.

[27] Donisi, D., B. Bellini, R. Beccherelli, R. Asquini, G. Gilardi, M. Trotta and A. d' Alessandro. 2010. A switchable liquid-crystal optical channel waveguide on silicon. *IEEE J. Quant. Electron.* 46: 762–768.

[28] Needleman, D. and Z. a d Dogic. 2017. Active matter at the interface between materials science and cell biology. *Nature Reviews Materials* 2: 17048.

[29] Dabrowski, R., K. Garbat, S. Urban, T.R. Wolinski, J. Dziaduszek, T. Ogrodnik and A. Siarkowska. 2017. Liquid crystals, pp. 1–18.

[30] Popov, N., L.W. Honaker, M. Popov, N. Usol'tseva, E.K. Mann, A. J´akli and P. Popov. 2018. Thermotropic liquid crystal-assisted chemical and biological sensors. *Materials* 11(1): 20.

[31] Popov, P. 2015. *Liquid Crystal Interfaces: Experiments, Simulations and Biosensors.* KentState University. https://etd.ohiolink.edu/apexprod/rws_olink/r/1501/10?clear=10&p10_accession_num=kent1434926908.

[32] Popov, P., E.K. Mann and A. J´akli. 2017. Thermotropic liquid crystal films for biosensors and beyond. *J. Mater. Chem. B* 5(26): 5061–5078.

[33] Eimura, H., D.S. Miller, X. Wang, N.L. Abbott and T. Kato. 2016. Self-assembly of bioconjugated amphiphilic mesogens having specific binding moieties at aqueous-liquid crystal interfaces. *Chemistry of Materials* 28(4): 1170–1178.

[34] Szilv´asi, T., L.T. Roling, H. Yu, P. Rai, S. Choi, R.J. Twieg, M. Mavrikakis and N.L. Abbott. 2017. Design of chemoresponsive liquid crystals through integration of computational chemistry and experimental studies. *Chem. Mater.* 29(8): 3563–3571.

[35] Liu, Q., F. Zuo, Z. Zhao, J. Chen and D. Xu. 2017. Molecular dynamics investigations of an indicator displacement assay mechanism in a liquid crystal sensor. *Phys. Chem. Chem. Phys.* 19(35): 23924–23933.

[36] Kaushik, A.K. and K.D. Chandra. 2016. *Nanobiotechnology for Sensing Applications: from Lab to Field*, first ed. Apple Academic Press, Canada.

[37] Joshi, T., A. Kumar, J. Prakash and A.M. Biradar. 2010. Low power operation of ferroelectric liquid crystal system dispersed with zinc oxide nanoparticles. *Appl. Phys. Lett.* 96(25): 253109.

[38] Lapanik, A., A. Rudzki, B. Kinkead, H. Qi, T. Hegmann and W. Haase. 2012. Electrooptical and dielectric properties of alkylthiol-capped gold nanoparticle–ferroelectric liquid crystal nanocomposites: influence of chain length and tethered liquid crystal functional groups. *Soft Matter* 8(33): 8722.

[39] Lisetski, L.N., S.S. Minenko, A.V. Zhukov, P.P. Shtifanyuk and N.I. Lebovka. 2009. Dispersions of carbon nanotubes in cholesteric liquid crystals. *Molecular Crystals and Liquid Crystals* 510(1): 43–50.

[40] Li, F., O. Buchnev, C.I. Cheon, A. Glushchenko, V. Reshetnyak, Y. Reznikov, T.J. Sluckin and J.L. West. 2006. Orientational coupling amplification in ferroelectric nematic colloids. *Phys. Rev. Lett.* 97(14).

[41] Kato, T., N. Mizoshita and K. Kishimoto. 2006. Functional liquid-crystalline assemblies: Self-organized soft materials. *Angew. Chem. Int. Ed.* 45(1): 38–68.

[42] Shen, Y. and I. Dierking. 2019. Perspectives in liquid-crystal-aided nanotechnology and nanoscience. *Appl. Sci.* 9(12): 2512.

[43] Hartono, D., W.J. Qin, K.-L. Yang and L.-Y.L. Yung. 2009. Imaging the disruption of phospholipid monolayer by protein-coated nanoparticles using ordering transitions of liquid crystals. *Biomaterials* 30(5): 843–849.

[44] Wei, Y. and C.-H. Jang. 2017. Visualization of cholylglycine hydrolase activities through nickel nanoparticle-assisted liquid crystal cells. *Sensor. Actuator. B Chem.* 239: 1268–1274.

[45] Kaushik, A., R.D. Jayant, S. Tiwari, A. Vashist and M. Nair. 2016a. Nano-biosensors to detect beta-amyloid for Alzheimer's disease management. *Biosens. Bioelectron.* 80: 273–287.

[46] Kaushik, A., R. Kumar, E. Huey, S. Bhansali, N. Nair and M. Nair. 2014. Silica nanowires: Growth, integration, and sensing applications. *Microchimica Acta* 181(15–16): 1759–1780.

[47] Kaushik, A. and M. Mujawar. 2018. Point of care sensing devices: Better care for everyone. *Sensors* 18(12): 4303.

[48] Choudhary, A., T. George and G. Li. 2018. Conjugation of nanomaterials and nematic liquid crystals for futuristic applications and biosensors. *Biosensors* 8(3): 69.

[49] Vallamkondu, J., E. Corgiat, G. Buchaiah, R. Kandimalla and P. Reddy. 2018. Liquid crystals: A novel approach for cancer detection and treatment. *Cancers* 10(11): 462.

[50] Funari, R., B. Della Ventura, R. Carrieri, L. Morra, E. Lahoz, F. Gesuele, C. Altucci and R. Velotta. 2015. Detection of parathion and patulin by quartz-crystal microbalance functionalized by the photonics immobilization technique. *Biosens. Bioelectron.* 67: 224–229.

[51] Kim, D.W., Y.H. Kim, H.S. Jeong and H.-T. Jung. 2011. Direct visualization of large-area graphene domains and boundaries by optical birefringency. *Nat. Nanotechnol.* 7(1): 29–34.

[52] Liao, S., Y. Qiao, W. Han, Z. Xie, Z. Wu, G. Shen and R. Yu. 2012. Acetylcholinesterase liquid crystal biosensor based on modulated growth of gold nanoparticles for amplified detection of acetylcholine and inhibitor. *Anal. Chem.* 84(1): 45–49.

[53] Liu, J. and Y. Lu. 2003. A colorimetric lead biosensor using DNAzyme-directed assembly of gold nanoparticles. *J. Am. Chem. Soc.* 125(22): 6642–6643.

[54] Yang, S., Y. Liu, H. Tan, C. Wu, Z. Wu, G. Shen and R. Yu. 2012. Gold nanoparticle based signal enhancement liquid crystal biosensors for DNA hybridization assays. *Chemical Communications* 48(23): 2861.

[55] Li, X., G. Li, M. Yang, L.-C. Chen and X.-L. Xiong. 2015. Gold nanoparticle based signal enhancement liquid crystal biosensors for tyrosine assays. *Sensor. Actuator. B Chem.* 215: 152–158.

[56] Roufs, J.B. 1990. L-tyrosine in the treatment of narcolepsy. *Med. Hypotheses* 33(4): 269–273.

[57] McGeer, P.L. and E.G. McGeer. 1973. Neurotransmitter synthetic enzymes. *Progress in Neurobiology* 2: 69–117.

[58] Maragoudakis, M.E. and N.E. Tsopanoglou. 2019. *Thrombin: Physiology and Disease*, first ed. Springer, New York.

[59] McCamley, M.K., A.W. Artenstein and G.P. Crawford. 2007a. Liquid crystal biosensors. pp. 241–296. *In*: Woltman, S.J., G.P. Crawford and G.D. Jay (eds.). *Liquid Crystals: Frontiers in Biomedical Applications*. World Scientific Publishing, Singapore.

[60] McCamley, M.K., A.W. Artenstein, S.M. Opal and G.P. Crawford. 2007b. Optical detection of sepsis markers using liquid crystal based biosensors. *Proc. SPIE* 6441: 64411Y(1–8).

[61] Saeed, M.H. et al. 2020. Recent Advances in the polymer dispersed liquid crystal composite and its applications. *Molecules* 25(23): 5510. doi:10.3390/molecules25235510.

[62] Ji, Y.-Y. et al. 2020. Terahertz birefringence anisotropy and relaxation effects in polymer-dispersed liquid crystal doped with gold nanoparticles. *Opt. Express.* 28(12): 17253. doi:10.1364/OE.392773.

[63] Jayoti, D. and P. Malik. 2018. Dielectric study of gold nanoparticle doped polymer dispersed liquid crystal. AIP Conference Proceedings (1953) 100085.

[64] Kanayama, N. et al. 2001. Distinct thermodynamic behaviour of a mesomorphic gold nanoparticle covered with a liquid-crystalline compound. *Chem. Commun.* (24): 2640. doi:10.1039/b108909a.

[65] Qi, H. and T. Hegmann. 2006. Formation of periodic stripe patterns in nematic liquid crystals doped with functionalized gold nanoparticles. *J. Mater. Chem.* 16(43): 4197. doi:10.1039/b611501b.

[66] Qi, H., J.O. Neil and T. Hegmann. 2008. Chirality transfer in nematic liquid crystals doped with (S)-naproxen-functionalized gold nanoclusters: An induced circular dichroism study. *J. Mater. Chem.* 18(4): 374. doi:10.1039/B712616F.

[67] Podgornov, F.V., A.V. Ryzhkova and W. Haase. 2010. Influence of gold nanorods size on electro-optical and dielectric properties of ferroelectric liquid crystals. *Appl. Phys. Lett.* 97(21): 212903. doi:10.1063/1.3517486.

[68] Pratibha, R., W. Park and I.I. Smalyukh. 2010. Colloidal gold nanosphere dispersions in smectic liquid crystals and thin nanoparticle-decorated smectic films. *J. Appl. Phys.* 107(6): 063511. doi:10.1063/1.3330678.

[69] Khatua, S. et al. 2010. Plasmonic nanoparticles–liquid crystal composites. *J. Phys. Chem. C.* 114(16): 7251. doi:10.1021/jp907923v.

[70] Inam, M. et al. 2011. Effect of gold nano-particles on switch-on voltage and relaxation frequency of nematic liquid crystal cell. *AIP Adv.* 1(4): 042162. doi:10.1063/1.3668125.

[71] Krishna Prasad, S. et al. 2006. Electrical conductivity and dielectric constant measurements of liquid crystal–gold nanoparticle composites. *Liq. Cryst.* 33(10): 1121. doi:10.1080/02678290600930980.

[72] Koenig, G.M. Jr et al. 2007. Coupling of the plasmon resonances of chemically functionalized gold nanoparticles to local order in thermotropic liquid crystals. *Chem. Mater.* 19(5): 1053. doi:10.1021/cm062438p.

[73] Cseh, L. and G.H. Mehl. 2007. Structure–property relationships in nematic gold nanoparticles. *J. Mater. Chem.* 17(4): 311. doi:10.1039/B614046G.

[74] Cseh, L. and G. Mehl. 2006. The design and investigation of room temperature thermotropic nematic gold nanoparticles. *J. Am. Chem. Soc.* 128(41): 13376. doi:10.1021/ja066099c.

[75] Qi, H., B. Kinkead and T. Hegmann. 2008. Unprecedented dual alignment mode and freedericksz transition in planar nematic liquid crystal cells doped with gold nanoclusters alline gold metamaterials. *Adv. Funct. Mater.* 18(2): 212. doi:10.1002/adfm.200701327.

[76] Qi, H. et al. 2009. Miscibility and alignment effects of mixed monolayer cyanobiphenyl liquid-crystal-capped gold nanoparticles in nematic cyanobiphenyl liquid crystal hosts. *Chemphyschem.* 10(8): 1211. doi:10.1002/cphc.200800765.

[77] Qi, H. and T. Hegmann. 2009. Multiple alignment modes for nematic liquid crystals doped with alkylthiol-capped gold nanoparticle. *ACS Appl. Mater. Interfaces* 1(8): 1731. VOL. NO. doi:10.1021/am9002815.

[78] Sridevi, S. et al. 2010. Enhancement of anisotropic conductivity, elastic, and dielectric constants in a liquid crystal-gold nanorod system. *Appl. Phys. Lett.* 97(15): 151913. doi:10.1063/1.3499744.

[79] Yoshida, H. et al. 2010. Nanoparticle-dispersed liquid crystals fabricated by sputter doping. *Adv. Mater.* 22(5): 622. doi:10.1002/adma.200902831.

[80] Urbanski, M. et al. 2010. Director field of birefringent stripes in liquid crystal/nanoparticle dispersions. *Liq. Cryst.* 37(9): 1151. doi:10.1080/02678292.2010.489160.

[81] Qi, H. and T. Hegmann. 2011. Liquid crystal–gold nanoparticle composites. *Liquid Crystals Today* 20(4): 102. doi:10.1080/1358314X.2011.610133.

[82] Draper, M. et al. 2011. Self-assembly and shape morphology of liquid crystalline gold metamaterials. *Adv. Funct. Mater.* 21(7): 1260. doi:10.1002/adfm.201001606.

[83] Milette, J. et al. 2011. Self-assembly and shape morphology of liquid crystalline gold metamaterials. *J. Mater. Chem.* 21(25): 9043. doi:10.1039/c1jm10553a.

[84] Vardanyan, K.K., E.D. Palazzo and R.D. Walton. 2011. Liquid crystal composites with a high percentage of gold nanoparticles. *Liq. Cryst.* 38(6): 709. doi:10.1080/02678292.2011.569760.

[85] Vardanyan, K.K., R.D. Walton and D.M. Bubb. 2011. Liquid crystal composites with a high percentage of gold nanoparticles. *Liq. Cryst.* 38(10): 1279. doi:10.1080/02678292.2011.610469.

[86] Vardanyan, K.K. et al. 2012. Liquid crystalline cyanobiphenyl homologues doped with gold nanoparticles. *Liq. Cryst.* 39(9): 1083. doi:10.1080/0267829 2.2012.696729.

[87] Vardanyan, K.K. et al. 2012. Study of pentyl-cyanobiphenyl nematic doped with gold nanoparticles. *Liq. Cryst.* 39(5): 595. doi:10.1080/02678292.2012.6 68567.

[88] Montazami, R. et al. 2012. Enhanced thermomechanical properties of a nematic liquid crystal elastomer doped with gold nanoparticles. *Sens. Actuators A* 178: 175. doi:10.1016/j.sna.2012.01.026.

[89] Mang, X. et al. 2012. Control of anisotropic self-assembly of gold nanoparticles coated with mesogens. *J. Mater. Chem.* 22(22): 11101. doi:10.1039/c2jm16794h.

[90] Podoliak, N. et al. 2012. High optical nonlinearity of nematic liquid crystals doped with gold nanoparticles. *J. Phys. Chem. C.* 116(23): 12934. doi:10.1021/ jp302558c.

[91] Prasad, S.K. et al. 2014. Enhancement of electrical conductivity, dielectric anisotropy and director relaxation frequency in composites of gold nanoparticle and a weakly polar nematic liquid crystal. *RSC Adv.* 4(9): 4453. doi:10.1039/C3RA45761C.

[92] Choudhary, A. and Guoqiang, L.I. 2014. Anisotropic shift of surface plasmon resonance of gold nanoparticles doped in nematic liquid crystal. *Opt. Express.* 22(20): 24348. doi:10.1364/OE.22.024348.

[93] Mirzaei, J. et al. Synthesis of liquid crystal silane-functionalized gold nanoparticles and their effects on the optical and electro-optic properties of a structurally related nematic liquid crystal. *Chem. Phys. Chem.* 15(7): 1381. doi:10.1002/cphc.201301052.

[94] Urbanski, M. et al. 2014. Synthesis of liquid crystal silane-functionalized gold nanoparticles and their effects on the optical and electro-optic properties of a structurally related nematic liquid crystal. *Chem. Phys. Chem.* 15(7): 1395. doi:10.1002/cphc.201301054.

[95] Vardanyan, K.K. et al. 2015. Multicomponent nematic systems with doped gold nanoparticles. *Liq. Cryst.* 42(4): 445. doi:10.1080/02678292.2014.996793.

[96] Xue, C. et al. 2015. Light-driven reversible alignment switching of liquid crystals enabled by Azo thiol grafted gold nanoparticles. *Chemphyschem.* 16(9): 1852. doi:10.1002/cphc.201500194.

[97] Mori, T., A. Sharma and T. Hegmann. 2016. Significant enhancement of the chiral correlation length in nematic liquid crystals by gold nanoparticle surfaces featuring axially chiral binaphthyl ligands. *ACS Nano.* 10(1): 1552. doi:10.1021/acsnano.5b07164.

[98] Hsu, C.-J. et al. 2017. Electro-optical effect of gold nanoparticle dispersed in nematic liquid crystals. *Crystals* 7(10): 287. doi:10.3390/cryst7100287.

[99] Budaszewski, D., M. Chychlowski et al. 2019. Enhanced efficiency of electric field tunability in photonic liquid crystal fibers doped with gold nanoparticles. *Opt. Express.* 27(10): 14260. doi:10.1364/OE.27.014260.

[100] Tomczyk, E. et al. 2019. Gold nanoparticles thin films with thermo- and photoresponsive plasmonic properties realized with liquid-crystalline ligands. *Small* 15(37): 1902807. doi:10.1002/smll.201902807.

[101] Yoshida, H. et al. 2009. Nanoparticle-stabilized cholesteric blue phases. *Appl. Phys. Express* 2(12): 121501. doi:10.1143/APEX.2.121501.

[102] Yabu, S. et al. 2011. Polarization-independent refractive index tuning using gold nanoparticle-stabilized blue phase liquid crystals. *Opt. Lett.* 36(18): 3578. doi:10.1364/OL.36.003578.

[103] Wong, J.M., J.Y. Hwang and L.C. Chien. 2011. Electrically reconfigurable and thermally sensitive optical properties of gold nanorods dispersed liquid crystal blue phase. *Soft Matter* 7(18): 7956. doi:10.1039/c1sm05764b.

[104] Yoshida, H. et al. 2013. Phase-dependence of gold nanoparticle dispersibility in blue phase and chiral nematic liquid crystals. *Opt. Mater. Express* 3(6): 842. doi:10.1364/OME.3.000842.

[105] Liu, Y.-S., H.-C. Lin and H.-L. Xu. 2018. The surface plasmon resonance effect on the defect-mode cholesteric liquid crystals doped with gold nanoparticles. *IEEE Photonics J.* 10(5): 1. doi:10.1109/JPHOT.2018.2821560.

[106] Kaur, S. et al. 2007. Enhanced electro-optical properties in gold nanoparticles doped ferroelectric liquid crystals. *Appl. Phys. Lett.* 91(2): 023120. doi:10.1063/1.2756136.

[107] Prakash, J. et al. 2008. Nonvolatile memory effect based on gold nanoparticles doped ferroelectric liquid crystal. *Appl. Phys. Lett.* 93(11): 112904. doi:10.1063/1.2980037.

[108] Kumar, A. et al. 2009. Enhanced photoluminescence in gold nanoparticles doped ferroelectric liquid crystals. *Appl. Phys. Lett.* 95(2): 023117. doi:10.1063/1.3179577.

[109] Pratibha, R. et al. 2009. Tunable optical metamaterial based on liquid crystal-gold nanosphere composite. *Opt. Express.* 17(22): 19459. doi:10.1364/OE.17.019459.

[110] Podgornov, F.V., A.V. Ryzhkova and W. Haase. 2010. Influence of gold nanorods size on electro-optical and dielectric properties of ferroelectric liquid crystals. *Appl. Phys. Lett.* 97(21): 212903. doi:10.1063/1.3517486.

[111] Mirzaei, J. et al. 2013. Hydrophobic gold nanoparticles via silane conjugation: chemically and thermally robust nanoparticles as dopants for nematic liquid crystals. *Philos. Trans. A. Math. Phys. Eng. Sci.* 371(1988): 20120256. doi: 10.1098/rsta.2012.0256.

[112] Lapanik, A. et al. 2012. Electrooptical and dielectric properties of alkylthiol-capped gold nanoparticle–ferroelectric liquid crystal nanocomposites: influence of chain length and tethered liquid crystal functional groups. *Soft Matter* 8(33): 8722. doi:10.1039/c2sm25991e.

[113] Hinojosa, A. and S.C. Sharma. 2010. Effects of gold nanoparticles on electro-optical properties of a polymer-dispersed liquid crystal. *Appl. Phys. Lett.* 97(8): 081114. doi:10.1063/1.3482942.

[114] Wang, K. et al. 2015. Improvement on the performance of holographic polymer-dispersed liquid crystal gratings with surface plasmon resonance of Ag and Au nanoparticles. *Plasmonics* 10(2): 383. doi:10.1007/s11468-014-9819-8.

[115] Ji, Y.-Y. et al. 2020. Terahertz birefringence anisotropy and relaxation effects in polymer-dispersed liquid crystal doped with gold nanoparticles. *Opt. Express.* 28(12): 17253. doi:10.1364/OE.392773.

[116] Siarkowska, A. et al. 2017. Thermo- and electro-optical properties of photonic liquid crystal fibers doped with gold nanoparticles. *Beilstein J. Nanotechnol.* 8: 2790. doi:10.3762/bjnano.8.278.

CHAPTER 4
Liquid Crystal Nanoparticles in Commercial Drug Delivery System

1. Introduction

Continuous improvement in drug efficacy has always been the target of nanoparticle medicated drug delivery [1–4]. Basic benefits of using nanoparticles in drug delivery are that conventional drug delivery issues like insolubility and toxicity are removed [15–16]. Different techniques and methods are being developed so that new nanoparticles can interact with targeted cells [17–18]. But limitation is the production of these nanoparticles on a large scale. Liquid crystal nanoparticles have been used in drug delivery, as they are better than conventional nanoparticles. Usually liquid crystal nanoparticles have three types of components; liquid crystal cross-linking agent, homologue of the organic chromophore perylene, and polymerizable surfactant containing a carboxylate headgroup [5, 6]. Mini emulsion technique is used to thermally polymerize these components and create stable NP as a colloidal solution [19–20]. LCNPS have been synthesized on the basis of their requirement, by changing the concentration of the two components. On varying the liquid crystal cross-linking agent and homologue of the organic chromophore perylene ratio, photo and chemical stability can be increased so that LCNPS could be used in fluorescence-biological applications.

2. Importance of Liquid Crystals as Nanomaterial

Many drugs have potential to treat cancer, respiratory diseases, pulmonary diseases and have anti-oxidant, anti-inflammatory, anti-bacterial, anti-diabetic properties [7]. But they have also some limited properties like; poor solubility, bioavailability, chemical fragility, hydrophilic nature, hydrophobic, poor intestinal absorption, dose-dependent cardiotoxicity and bioavailability. These limitations reduce a lot due to drug efficacy. Table 4.1 is showing some active ingredient/drugs, their use in different diseases and their limited properties.

But it has been seen that liquid crystalline nanoparticulate can improve solubility and play an important role in controlling the drug release rate. Lipid based systems and self-emulsifying drug delivery systems are much common in liquid crystalline systems but their 2D and 3D periodic structural arrangement are different [21].

Commonly, liquid crystals can be divided in two category; lyotropic and thermotropic. Thermotropic LC change their properties with temperature. But lyotropic are affected by their amphiphilic molecules that are much affected by solvent (aqueous mostly). Lyotropic are not degraded by enzymes or hydrolysed in releasing drug, so reach the target. There is also good absorption of lyotropic loaded drugs by human tissues. The whole system is decided by amphiphilic molecules, pH, temperature, medium, pressure and salt concentrations. This amphiphilic nature of molecules used in LC, hydrophilic, lipophilic and amphiphilic molecules of drugs can be used to be encapsulated [21–25]. In most of the cases, two forms of lyotropic liquid crystals are used in drug delivery, first are cubosomes and second are hexosomes. These two LCNP are of high interest in the drug delivery field due extraordinary potential as drug vehicles [26–27]. Non-toxic, biodegradable and bioadhesive characteristics are some of the properties of these LCNPS [28–29]. Lyotropic LCNP surface molecules have the property that they absorbs water from environment. Depending on the absorbance of water from environment, it may convert to any of the phases; laminar, cubic or hexagonal. Three factors that can be affecting phase transition are:

Table 4.1. Application of active ingredient/drugs, in different diseases and their properties.

S. no.	Active Ingredient/Drug	Application	Limited property	References
1.	Quercetin (found in plant)	anti-oxidant, anti-inflammatory, anti-bacterial, anti-diabetic and anti-cancer	Hydrophobic nature	7
2.	Fisetin(found in *Cotinus coggygria* tree)	anti-oxidant, anti-inflammatory, hypoglycaemic and anti-cancer.	Hydrophobic nature	8
3.	*Brucea javanica* (*herbal medicine*)	various cancer treatment	dose-dependent cardiotoxicity	9
4.	Doxorubicin	Cancer treatment	dose-dependent cardiotoxicity	10
5.	Polyphenols (obtained from *Cornus mas L.*)	Antioxidant, anti-cancer and anti-inflammatory	degrade in gastrointestinal	11
6.	Curcumin	Therapeutic application	poor solubility, bioavailability	12
7.	Crocins	Respiratory diseases	Hydrophilic nature	13
8.	Gambogenic acid	Cancer	Less bioavailability	14

repulsion force between head groups of interfaces of surfactant and the water, degree of contact between water and alkyl group and conformational disorder in alkyl chain. Hexagonal mesophase are formed in two ways: normal and reverse mesophase, on the basis of two solvents; aqueous and organic respectively. In organic solvents or anhydrous solvents, polar groups shift towards the inner side and get covered by non polar hydrocarbon chains. These are called reverse hexagonal mesophase. Cubic form is formed in excess of water contents. Release of drug from the cubic phase is controlled because of its packed microstructure. Bioadhesive property of cubic phase structure makes cubic lyotropic LC NP suitable for *in vitro* studies. This packed structure is suitable for the pulmonary, vaginal, buccal and nasal drug delivery [28].

3. Different Methods and Techniques used for the Synthesis and Characterization of Liquid Crystal Mesophases/Nanoparticles

There are different methods for synthesizing LCNPs/cubosomes and hexosomes. Here are some of the methods that have been adopted by many scientists.

1. Miniemulsion method

Miniemulsion method is the simplest method that is used by many scientists for synthesis of LCNPs. This method provides chemically stable and early synthesis of LCNPs. For any type of formation of LCNPs, LCNPs are made of three components; perylene-based dye, liquid crystal cross-linking agent and carboxyl-terminated polymerizable surfactant. Polyacrylate nanoparticle primarily composed of (cross-linking agent) DACTP11 and capped with (surfactant) AC10COONa have been synthesized by miniemulsion process. First cross-linking agent solution was made in chloroform. Then dye was also dissolved in this solution. This mixture was then added in water with continuous stirring and sonicating for getting the micelles. Then the solution was heated to start the polymerization for getting permanent micelles in the form of nanoparticles or simply LCNPs suspended colloidal solution [17].

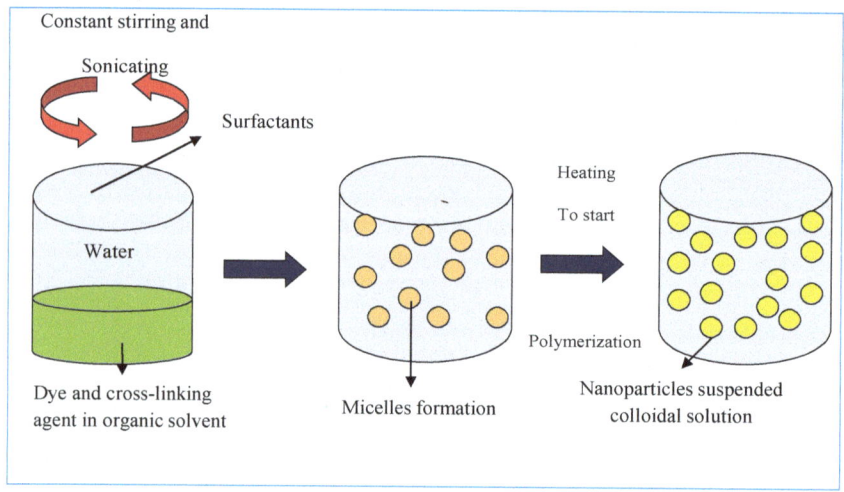

Figure 4.1. Diagrammatical representation of synthesis of LCNPs by miniemulsion method.

Diagrammatical representations of synthesis of LCNPs are shown in given Figure 4.1.

2. Top-down preparation

Top-down method is the oldest and most popular synthetic methodology of cubosomes and hexosomes, which was initiated in nineteen's century. First a homogeneous solution of amphiphilic molecules + drug + polymeric stabilizer is prepared by using sonication or high pressure homogenizer. In this method, stabilizer and temperature play an important role for the synthesis of nanoparticles. In most of the cases, pluronic F127, as stabilizer polymer has been used [42–44]. But this, method has a drawback that large energy is required in making homogeneous mixture and also the product has some impurities of vesicles. This is represented in Figure 4.2.

3. Bottom-up preparation

Mostly cubosomes are made by this method. Three important components are required for this preparation; amphiphiles molecules, drug and hydrotropic molecules. Hydrotropes are the compounds which solubilise hydrophobic compounds (similar to

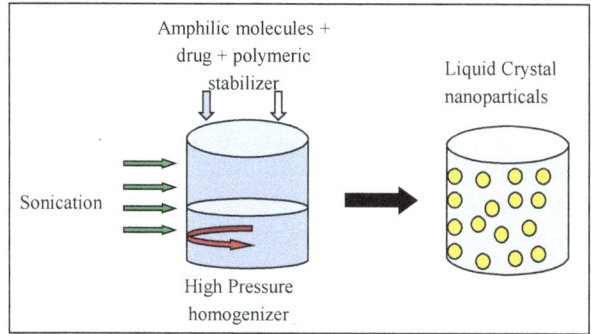

Figure 4.2. Synthesis of LCNPS by Top-down method.

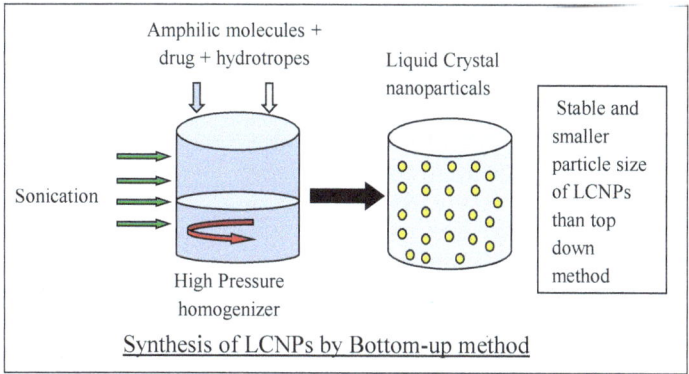

Figure 4.3. Synthesis of LCNPS by Bottom-up method.

surfactants). But in hydrotropes, hydrophobic part is very small and it is basically used for increasing solubility of poorly soluble solutes and prevent aggregation [45–46]. Hydrotropes save the energy required to get the homogeneous solution as in case of top-down method. Another benefit of this method is the production of lesser size cubosomes with greater stability. But there are also draw backs of this method. These hydrotropes may cause some side effects within the human body and second is that vesicles are also formed in this method. The schematic diagram of its synthetic strategy is represented in Figure 4.3.

4. Heating method

Heating method has benefits over above mentioned two methods. First vesicles are totally removed from the solution and no extra hydrotropes are added. This results in increase in no of cubosomes with more stability [47]. Thus reduction in vesicles increases the drug release efficacy by cubosomes. Basically heating decreases the surfactant solubility, so vesicle come closer and more cubosomes are formed [48]. But the biggest drawback with this method is that this method can be applied only above hundred degrees centigrade, and most of the proteins and DNA decompose at this temperature.

5. Spray drying

This is an old and famous method since 1990 for the formation of cubosomes and hexosomes [40–41]. In this method, precursor is prepared first by combination of organic solvent/water solvent and other stabilizer/carrier and then they are spray dried to obtain powder form. Aqueous cubosomes are made by this process and then reduced to powder form. Rehydration of this powder gives back aqueous form of the liquid crystal in dispersed form. Spray drying method can be used for medicines that can be used in inhalers. Scientists have found that such type of powder inhaling medicines are fast and more effective than conventional medicine/tablets [49].

4. Applications of Liquid Crystal NPs in Drug Delivery

Lytropic LCNPs are being used in drug delivery because they have the following abilities.

1. Sustained or controlled drug release.
2. Improved drug bioavailability and reduced drug toxicity.
3. Enhanced stability of drugs.
4. Increased penetration of drugs.
5. Stimuli responsive drug delivery system.

All types of molecules like lipophilic and hydrophilic can be used, included in liquid crystals this is the most important property of liquid crystal to be used in drug delivery [30]. All three phases of liquid crystals; lamellenar, cubic or hexagonal mesophase have been used in drug delivery.

1. Lamellar phases

Arsphenamin medicine used in syphilis was initially used with liquid crystals lamellar phases [31]. This medicine is also used as antimicrobial. Lamellar phases have much similar structure to skin, i.e., it is used in controlled drug released. Lamellar phase has double layer of surfactant having polar head groups which are expended between two aqueous interfaces. Relative fluidity is also another property of lamellar phase.

2. Cubic phases

The cubic mesophases are of interest in the creation of cubosomes, as the water channels have extremely high surface area (up to 400 m^2/g) available for the loading of drugs. Reverse cubic phase is commonly used for production of liquid crystal nanoparticles.

3. Hexagonal phase

This is most abundant and commonly forming lyotropic phase. Hexagonal phase is denoted by H_1 phase. A number of phases have been used to make liquid crystals nanoparticles. Cubosomes have been synthesized using both topology mesophase; normal and inverse. But hexosomes are made only by using inverse topological mesophases [39]. Reverse hexagonal phase is similar to normal hexagonal phase and this is less complex than reverse cubic phase. Reverse hexagonal phase requires higher amphiphilic concentration than in reverse cubic phase. Hexagonal packing makes hexosomes more stable, due to less spacing between the tails of amphiphilic molecules.

The following table is shows different drug loaded lytropic liquid crystal nanoparticles in different applications:

S. No.	Drug loaded lytropic LC NPs	Application	Reference no.
1.	Hydrophobic peptide-based drugs loaded LCNPs	Five-fold enhancement of bioavailability, sustained release for liver based drugs.	53
2.	Topical drug delivery based on cubosomes and hexosomes	Good adsorption for skin	50
3.	Carbomer-indomethacin loaded LCNPs	UVB-induced erythema	51
4.	Progesterone loaded hexosomes	Enhanced transmucosal flux more than five fold	52
5.	Photosensitizer Chlorin activated nanoparticles in animal models	Photodynamic therapy-higher drug	54
6.	Cyclosporine A-loaded GMO/F127 cubosome	Reduced ocular irritancy; also may used in cornea, conjunctiva, oriris treatment	55
7.	OG-based hexosomes	Irinotecan (anticancer drug) stability increases	56
8.	GMO-based hexosomes	Vitamin K delivery to the *stratum corneum* increased 2 times	57
9.	Aspirin and vitamin E loaded cubic LCNPs	Sustained release	58
10	Tetracycline, timolol maleate, chlorpheniramine maleate, propranolol hydrochloride, melatonin, pindolol, propranolol and pyrimethamine and hemoglobin loaded cubic LCNPs	Sustained release	59–64

5. Factors Affecting Drug Release/Delivery by Lyotropic Liquid Crystals

Lyotropic liquid crystals are soft materials and are much affected with change in surroundings/medium changes like; temperature, pH, pressure, etc. On increasing the pH of the medium, degree of ionization of head group may increase. This increases the repulsion between the negatively charge head groups and the surface area

124

increases. This loss the packing of lyotropic LC phases and transition of the phase may be started. This feature/effect of pH on packing of lyotropic LC phases may be used for taregetted drug delivery. As we know that in the human body, different parts have different pH values. Scientists have been working on release of drug with different lytropic LC phases at different pH ranges. As intestine pH is 7 and our stomach is 2. Thus this study has been utilizing for targeted drug delivery without any leakages [32–33].

Second important parameter that affects the drug delivery is temperature [34–37]. A temperature based study was carried out by researchers to see the effect of temperature on phase transition of lyotropic phases; phytantriol matrix with glucose loaded in (40 to 30)°C range. It was found a clear transition in phases from hexagonal phase to cubic phase. Release was found much faster in cubic phase.

Other parameters are also responsible for the LLC drug release like; drug size, viscosity solubility of the drug-hydrophobic or hydrophilic nature [38]. Surface area and the volume are much affected in lyotropic phases on the basis of type of hydrophilic and hydrophobic nature of the drug. If the drug is attached to the hydrophilic part of the lyotropic phase, then it increases the volume of that part. But if drug is hydrophobic, then it increases the surface area of the phase.

6. Conclusions

This chapter has mainly focused on the synthesis of LCNPs and their application in drug delivery. This chapter has described different possible methods for the synthesis of LCNPs. It also includes the factors affecting the drug delivery and how their draw backs can be removed by drug loaded LCNPs. A detailed summary for the drug delivery application of LCNPs has been discussed. Lyotropic based delivery has been found very effective due to many reasons like; lyotropic LCs have the tendency to phase transition and are much affected by solution condition. This is very suitable for the drug delivery in human fluid. pH is another important factor that helps in sustained drug delivery by using LCNPs. Temperature based technology controls the release of drug in the case of LCNPs drug delivery method; as variation in temperature has been seen responsible for phase transition from hexagonal to cubic phase. Cubic phase has been found much comfortable for drug release. This chapter will be very beneficial for those researchers and scientists who are doing work on LCNPs synthesis and their applications in drug delivery. A lot work has been done on drug delivery application by LCNPs but less work is done on the interaction between LCNPs and the drug and also on the drawbacks of LCNPs. Long term effect of LCNPs on human body is also not discussed. Yet it is challenging to do commercialization LCNPs in industry and clinical purposes. A lot research is required still for lipid-based LCNPs drug delivery.

References

[1] Mo, J., G. Milleret and M. Nagaraj. 2017. Liquid crystal nanoparticles for commercial drug delivery. *Liquid Crystals Reviews* 5(2): 69–85.

[2] Gautam Singhvi, Saswata Banerjee and Archana Khosa. 2018. Chapter 11 - Lyotropic liquid crystal nanoparticles: A novel improved lipidic drug delivery system. *Organic Materials as Smart Nanocarriers for Drug Delivery* 471–517.

[3] Rajak, P., L.K. Nath and B. Bhuyan. 2019. Liquid crystals: an approach in drug delivery. *Indian J. Pharm. Sci.* 81(1): 11–21.

[4] Abhishesh, K. Mehata*, Deepa Dehari, Amit Gupta, Dangali C. Rabin and Alim Miya. 2021. Multifunctional liquid crystal nanoparticles for cancer therapy. *Current Nanomaterials* 6(1). https://doi.org/10.2174/24054615066 66210118114851.

[5] Christopher, M. Spillmann, Jawad Naciri, W. Russ Algar, Igor L. Medintz and James B. Delehanty. 2014. Multifunctional liquid crystal nanoparticles for intracellular fluorescent imaging and drug delivery. *ACS Nano* 8(7): 6986–6997. DOI: 10.1021/nn501816z.

[6] Vrinder Pal Singh, Meenu Mehta, Saurabh Satija, Deep Shikha Sharma and Pardeep Kumar Sharma. 2020. Emerging trends of liquid crystalline nanoparticles drug delivery for pulmonary disorders. *European Journal of Molecular & Clinical Medicine* 7(7): 2558.

[7] D'Andrea, G. 2015. Quercetin: A flavonol with multifaceted therapeutic applications? *Fitoterapia* 106: 256–271. doi:10.1016/j.fitote.2015.09.018.

[8] Mehta, P., A. Pawar, K. Mahadik and C. Bothiraja. 2018. Emerging novel drug delivery strategies for bioactive flavonol fisetin in biomedicine. *Biomedicine & Pharmacotherapy* 106: 1282–1291.

[9] Li, Y., A. Angelova, F. Hu, V.M. Garamus, C. Peng, N. Li ... A. Zou. 2019. pH responsiveness of hexosomes and cubosomes for combined delivery of Brucea javanica oil and Doxorubicin. *Langmuir* 35(45): 14532–14542. doi:10.1021/acs.langmuir.9b02257.

[10] Makai, M., E. Csanyi, I. Dekany, Z. Nemeth and I. Eros. 2003. Structural properties of non-ionic surfactant/glycerol/paraffin lyotropic liquid crystals. *Colloid Polym. Sci.* 281(9): 839–844.

[11] Radbeh, Z., N. Asefi, H. Hamishehkar, L. Roufegarinejad and A. Pezeshki. 2020. Novel carriers ensuring enhanced anti-cancer activity of Cornus mas (cornelian cherry) bioactive compounds. *Biomed Pharmacother.* 125: 109906. doi:10.1016/j.biopha.2020.109906.

[12] Yong, D.O.C., S.R. Saker, R. Wadhwa, D.K. Chellappan, T. Madheswaran, J. Panneerselvam ... V. Pillay. 2019. Preparation, characterization and *in-vitro* efficacy of quercetin loaded liquid crystalline nanoparticles for the treatment of asthma. *Journal of Drug Delivery Science and Technology* 54: 101297.

[13] Mahesh, K.V., S.K. Singh and M. Gulati. 2014. A comparative study of top-down and bottom-up approaches for the preparation of nanosuspensions of glipizide. *Powder Technology* 256: 436–449.

[14] Luo, Q., T. Lin, C.Y. Zhang, T. Zhu, L. Wang, Z. Ji ... W. Chen. 2015. A novel glyceryl monoolein-bearing cubosomes for gambogenic acid: Preparation, cytotoxicity and intracellular uptake. *International Journal of Pharmaceutics* 493(1-2): 30–9.

[15] Delehanty, J.B., J.C. Breger, G.K. Boeneman, M.H. Stewart and I.L. Medintz. 2013. Controlling the actuation of therapeutic nanomaterials: enabling nanoparticle-mediated drug delivery. *Ther. Delivery* 11: 1411–1429.

[16] Sailor, M.J. and J.H. Park. 2012. Hybrid nanoparticles for detection and treatment of cancer. *Adv. Mater.* 24: 3779–3802.

[17] Christopher, M. Spillmann, Jawad Naciri, W. Russ Algar, Igor L. Medintz and James B. Delehanty. 2014. Multifunctional liquid crystal nanoparticles for intracellular fluorescent imaging and drug delivery. *ACS Nano*, XXX (XX).

[18] Verma, A. and F. Stellacci. 2010. Effect of surface properties on nanoparticle-cell interactions. *Small* 6: 12–21.

[19] Spillmann, C.M., J. Naciri, K.J. Wahl, Y.H. Garner, M.-S. Chen and B.R. Ratna. 2009. Role of surfactant in the stability of liquid crystal-based nanocolloids. *Langmuir* 25: 2419–2426.

[20] Spillmann, C.M., J. Naciri, G.P. Anderson, M.S. Chen and B.R. Ratna. 2009. Spectral tuning of organic nanocolloids by controlled molecular interactions. *ACS Nano* 3: 3214–3220.

[21] Andrew Otte, Bong-Kwan Soh, Gwangheum Yoon and Kinam Park. 2018. Liquid crystalline drug delivery vehicles for oral and IV/subcutaneous administration of poorly soluble (and soluble) drugs. *International Journal of Pharmaceutics* 539: 175–183.

[22] Dong, Y.D., I. Larson, T. Hanley and B.J. Boyd. 2006. Bulk and dispersed aqueous phase behavior of phytantriol: effect of vitamin E acetate and F127 polymer on liquid crystal nanostructure. *Langmuir* 22(23): 9512–9518.

[23] Qiu, H. and M. Caffrey. 2000. The phase diagram of the monoolein/water system: metastability and equilibrium aspects. *Biomaterials* 21: 223–234.

[24] Nguyen, T.H., T. Hanley, C.J.H. Porter, I. Larson and B.J. Boyd. 2010. Phytantriol and glyceryl monooleate cubic liquid crystalline phases as sustained-release oral drug delivery systems for poorly water-soluble drugs I. Phase beaviour in physiologicallyrelevant media. *J. Pharm. Pharmacol.* 62(7): 844–855.

[25] Negrini, R. and R. Mezzenga. 2012. Diffusion, molecular separation, and drug delivery from lipid mesophases with tunable water channels. *Langmuir* 28(47): 16455–16462.

[26] Li, Y., C. Dong, D. Cun, J. Liu, R. Xiang and L. Fang. 2016. Lamellar liquid crystal improves the skin retention of 3-o. *AAPS PharmSciTech* 17(3): 767–77.

[27] Calixto, G.M., J. Bernegossi, L.M. de Freitas, C.R. Fontana and M. Chorilli. 2016. Nanotechnology-based drug delivery systems for photodynamic therapy of cancer: a review. *Molecules* 21: 342–50.

[28] Rajak, P., L.K. Nath and B. Bhuyan. 2019. Liquid crystals: an approach in drug delivery. *Indian Journal of Pharmaceutical Science* 81(1): 11–21.

[29] Guo, C., J. Wang, F. Cao, R.J. Lee and G. Zhai. 2010. Lyotropic liquid crystal systems in drug delivery. *Drug Discov Today* 15(23-24): 1032–40.

[30] Malmsten, M. 2006. *Surfactants and Polymers in Drug Delivery*. New York: Marcel Dekker Inc, 116–212.

[31] Wahlgren, S., A.L. Lindstrom and S.E. Friberg. 1984. Liquid crystals as a potential ointment vehicle. *J. Pharm. Sci.* 73: 1484–6.

[32] Negrini, R. and R. Mezzenga. 2011. ph-Responsive lyotropic liquid crystals for controlled drug delivery. *Langmuir* 27(9): 5296–5303.

[33] Negrini, R., W.K. Fong, B.J. Boyd and R. Mezzenga. 2015. ph-Responsive lyotropic liquid crystals and their potential therapeutic role in cancer treatment. *Chem.Commun.* 51(30): 6671–6674.

[34] Fong, W.K., T. Hanley and B.J. Boyd. 2009. Stimuli responsive liquid crystals provide on-demand drug delivery *in vitro* and *in vivo*. *J. Control. Release* 135(3): 218–226.

[35] Yaghmur, A., P. Laggner, S. Zhang and M. Rappolt. 2007. Tuning curvature and stability of monoolein bilayers by designer lipid-like peptide surfactants. *PLOS One* 2(5): e479.

[36] Czeslik, C., R. Winter, G. Rapp and K. Bartels. 1995. Temperature-and pressure-dependent phase behavior of monoacylglycerides monoolein and monoelaidin. *Biophys. J.* 68(4): 1423–1429.

[37] Lendermann, J. and R. Winter. 2003. Interaction of cytochrome c with cubic monoolein mesophases at limited hydration conditions: The effects of concentration, temperature and pressure. *Phys. Chem. Chem. Phys.* 5(7): 1440–1450.

[38] Sallam, A.l., F.F. Hamudi and E.A. Khalil. 2015. Effect of ethylcellulose and propylene glycol on the controlled-release performance of glyceryl monooleate-mertronidazole periodontal gel. *Pharmaceutical Develop. Tech.* 20(2): 159–168.

[39] Spicer, P.T. 2005. Progress in liquid crystalline dispersions: Cubosomes. *Cur. Opin. Colloid Interfac.* 10(5-6): 274–279.

[40] Ljusberg-Wahren, H., L. Nyberg and K. Larsson. 1996. Dispersion of the cubic liquid crystalline phase—structure, preparation, and functionality aspects. *Chim. Oggi* 14: 40–43.

[41] Spicer, P.T. 2005. Cubosome processing industrial nanoparticle technology development. *Chem. Eng. Res. Des.* 83: 1283–1286.

[42] Sagalowicz, L., M.E. Leser, H.J. Watzke and M. Michel. 2006. Monoglyceride self-assembly structures as delivery vehicles. *Trends in Food Science & Technology* 17(5): 204–214.

[43] Dong, Y.D., I. Larson, T. Haneley and B.J. Boyd. 2006. Bulk and dispersed aqueous phase behaviour of phytantriol: effect of vitamin E acetate and F127 polymer on liquid crystal nanostructure. *Langmuir* 22(23): 9512–9518.

[44] Worle, G., M. Drechsler, M.H. Koch, B. Siekmann, K. Westesen and H. Bunjes. 2007. Influence of composition and preparation parameters on the properties of aqueous monoolein dispersions. *Int. J. Pharm.* 329: 150–157.

[45] Vividha Dhapte and P. Mehta. 2015. Advance in hydrotropic solutions: an updated review, St. Petersburg polytechnical University. *Journal: Physics and Mathematics* 424–435.

[46] Larsson, K. 1989. Cubic lipid-water phases: structures and biomembrane aspects. *J. Phys. Chem.* 93(21): 7304–14.

[47] Barauskas, J., M. Johnsson, F. Joabsson and F. Tiberg. 2005. Cubic phase nanoparticles (cubosome): principles for controlling size, structure, and stability. *Langmuir* 21: 2569–2577.

[48] Wörle, G., B. Siekmann, M.H.J. Koch and H. Bunjes. 2006. Transformation of vesicular into cubic nanoparticles by autoclaving of aqueous monoolein/ poloxamer dispersions. *Eur. J. Pharm. Sci.* 27: 44–53.

[49] Spicer, P.T., W.B. Small II, W.B. Small, M.L. Lynch and J.L. Burns. 2002. Dry powder precursors of cubic liquid crystalline nanoparticles (cubosomes). *J. Nanopart. Res.* 4: 297–311.

[50] Niu, M., Y. Lu, L. Hovgaard and W. Wu. 2011. Liposomes containing glycocholate as potential oral insulin delivery systems: preparation, *in vitro* characterization, and improved protection against enzymatic degradation. *In. J. Nanomedicine* 6: 1155–1166.

[51] Esposito, E., R. Cortesi, M. Drechsler, L. Paccamiccio, P. Mariani, C. Contado, E. Stellin, E. Menegatti, F. Bonina and C. Pugilia. 2005. Cubosome dispersions as delivery systems for percutaneous administration of indomethacin. *Pharmac. Research* 22(12): 2163–2173.

[52] Petrilli, R. et al. 2013. Nanoparticles of lyotropic liquid crystals: A novel strategy for the topical delivery of a chlorin derivative for photodynamic therapy of skin cancer. *Current Nanoscience* 9(4): 434–441.

[53] Calixto, G.F. et al. 2016. Nanotechnology-based drug delivery systems for photodynamic therapy of cancer: A review. *Molecules* 21: 342–350.

[54] Lee, D.R., J.S. Park, I.H. Bae, Y. Lee and B.M. Kim. 2016. Liquid crystal nanoparticle formulation as an oral drug delivery system for liver-specific distribution. *Int. J. Nanomedicine* 11: 853–871.

[55] Chen, Y., Y. Lu, Y. Zhong, Q. Wang, W. Wu and S. Gao. 2012. Ocular delivery of cyclosporine A based on glyceryl monooleate/poloxamer 407 liquid crystalline nanoparticles: preparation,characterization, *in vitro* corneal penetration and ocular irritation. *J. Drug Target* 20(10): 856–63.

[56] Boyd, B.J., D.V. Whittaker, S.M. Khoo and G. Davey. 2006. Hexosomes formed from glycerate surfactants-formulation as a colloidal carrier for irinotecan. *Int. J. Pharm.* 318(1-2): 154–62.

[57] Lopes, L.B., F.F.F. Speretta and M.V. Bentley. 2007. Enhancement of skin penetration of vitamin K using monoolein-based liquid crystalline systems. *Eur. J. Pharm. Sci.* 32(3): 209–15.

[58] Wyatt, D. and D. Dorschel. 1992. A cubic-phase delivery system composed of glyceryl monooleate and water for sustained release of water-soluble drugs. *Pharmaceutical Technology* 16(10): 116–130.

[59] Esposito, E., V. Carotta, A. Scabbia et al. 1996. Comparative analysis of tetracycline-containing dental gels: poloxamer- and monoglyceride-based formulations. *International Journal of Pharmaceutics* 142(1): 9–23.

[60] Lindell, K., J. Engblom, M. Jonstr"omer, A. Carlsson and S. Engstr"om. 1998. Influence of a charged phospholipid on the release pattern of timolol maleate from cubic liquid crystalline phases. *Progress in Colloid and Polymer Science* 108: 111–118.

[61] Chang, C.-M. and R. Bodmeier. 1997. Swelling of and drug release from monoglyceride-based drug delivery systems. *Journal of Pharmaceutical Sciences* 86(6): 747–752.

[62] Costa-Balogh, F.O., E. Sparr, J.J.S. Sousa and A.C. Pais. 2010. Drug release from lipid liquid crystalline phases: relation with phase behavior. *Drug Development and Industrial Pharmacy* 36(4): 470–481.

[63] Burrows, R., J.H. Collett and D. Attwood. 1994. The release of drugs from monoglyceride-water liquid crystalline phases. *International Journal of Pharmaceutics* 111(3): 283–293.

[64] Leslie, S.B., S. Puvvada, B.R. Ratna and A.S. Rudolph. 1996. Encapsulation of hemoglobin in a bicontinuous cubic phase lipid. *Biochimica et BiophysicaActa* 1285(2): 246–254.

Index

B

Biosensing 87
Bluc Phasc 48
Bottom-up preparation 120

C

Carbon Nanotubes 59
Carbon NPs 90
Cellulose 23
Chiral LCs 20
Chirality 19
Cholesterics 11
Columnar mesophases 13, 54
Cubic phases 123

D

DNA 23
Drug delivery 123

F

Ferroelectric Nanoparticles 35

G

Gold Nanorods 94, 98
Gold NPs 104, 106

H

Heating method 122
Hexagonal phase 123

I

Ionic LCs 13

L

Lamellar phases 123
LCD 93
Lyotropic 6, 123

M

Magnetic nanoparticles 40
Metal Nanoparticles 36
Metal Nanorods 55
Miniemulsion method 119

N

Nano Clays 48, 61
Nano Discs 48
Nano Rods 35, 98
Nano-particle Doping 48
Nematics 8, 52, 85

O

One-dimensional nanorods 42

P

PDLC systems 106

Q

Quantum dots 39
Quasi-Zero-Dimensional 41

S

Semiconductor Nanorods 57
Smectics 12, 53, 94, 95
Spray drying 122

T

Thermotropic 7
Top-down preparation 120

Z

Zero-dimensional NPs 37, 39

.